COLLEGE AVENUE
KENTFIELD, CA 94904

WITHDRAWN

# World History Series

# Twentieth Century Science

**Titles in the World History Series**

The Age of Augustus
The Age of Feudalism
The Age of Pericles
The American Frontier
The American Revolution
Ancient Greece
The Ancient Near East
Architecture
Aztec Civilization
The Black Death
The Byzantine Empire
Caesar's Conquest of Gaul
The California Gold Rush
The Chinese Cultural
    Revolution
The Conquest of Mexico
The Crusades
The Cuban Missile Crisis
The Cuban Revolution
The Early Middle Ages
Egypt of the Pharaohs
Elizabethan England
The End of the Cold War
The French and Indian War
The French Revolution
The Glorious Revolution
The Great Depression
Greek and Roman Theater

The History of Slavery
Hitler's Reich
The Hundred Years' War
The Inquisition
The Italian Renaissance
The Late Middle Ages
The Lewis and Clark
    Expedition
The Mexican Revolution
The Mexican War of
    Independence
Modern Japan
The Punic Wars
The Reformation
The Relocation of the
    North American Indian
The Roman Empire
The Roman Republic
The Russian Revolution
The Scientific Revolution
The Spread of Islam
Traditional Africa
Traditional Japan
The Travels of Marco Polo
Twentieth Century Science
The Wars of the Roses
The Watts Riot
Women's Suffrage

# Twentieth Century Science

by
Harry Henderson and Lisa Yount

Lucent Books, P.O. Box 289011, San Diego, CA 92198-9011

*To the scientists and citizens
of the twenty-first century*

Library of Congress Cataloging-in-Publication Data

Henderson, Harry, 1951–
 Twentieth-century science / by Harry Henderson and Lisa Yount.
  p. cm.—(World history series)
 Includes bibliographical references and index.
 Summary: Provides an overview of scientific developments that have had significant impact on our world, including atomic physics, genetic engineering, and computer technology.
 ISBN 1-56006-304-1
 1. Science—History—Juvenile literature. [1. Science—History.] I. Yount, Lisa. II. Title. III. Series.
Q126.4.H46   1997
509'.04—dc21                                          96-49420
                                                          CIP
                                                           AC

Copyright 1997 by Lucent Books, Inc., P.O. Box 289011, San Diego, California 92198-9011

Printed in the U.S.A.

No part of this book may be reproduced or used in any other form or by any other means, electrical, mechanical, or otherwise, including, but not limited to, photocopy, recording, or any information storage and retrieval system, without prior written permission from the publisher.

# Contents

| | |
|---|---|
| Foreword | 6 |
| Important Dates in the History of Twentieth-Century Science | 9 |

**INTRODUCTION**
*Science Shapes a Century* — 10

**CHAPTER 1**
*Inside the Atom* — 14

**CHAPTER 2**
*The Expanding Universe* — 26

**CHAPTER 3**
*The Code of Life* — 40

**CHAPTER 4**
*Magic Bullets* — 54

**CHAPTER 5**
*Rebuilding the Body* — 66

**CHAPTER 6**
*The Information Revolution* — 77

**CHAPTER 7**
*An Unpredictable Future* — 89

| | |
|---|---|
| Notes | 97 |
| Glossary | 99 |
| For Further Reading | 104 |
| Works Consulted | 105 |
| Index | 108 |
| Picture Credits | 112 |
| About the Authors | 112 |

# Foreword

Each year on the first day of school, nearly every history teacher faces the task of explaining why his or her students should study history. One logical answer to this question is that exploring what happened in our past explains how the things we often take for granted—our customs, ideas, and institutions—came to be. As statesman and historian Winston Churchill put it, "Every nation or group of nations has its own tale to tell. Knowledge of the trials and struggles is necessary to all who would comprehend the problems, perils, challenges, and opportunities which confront us today." Thus, a study of history puts modern ideas and institutions in perspective. For example, though the founders of the United States were talented and creative thinkers, they clearly did not invent the concept of democracy. Instead, they adapted some democratic ideas that had originated in ancient Greece and with which the Romans, the British, and others had experimented. An exploration of these cultures, then, reveals their very real connection to us through institutions that continue to shape our daily lives.

Another reason often given for studying history is the idea that lessons exist in the past from which contemporary societies can benefit and learn. This idea, although controversial, has always been an intriguing one for historians. Those that agree that society can benefit from the past often quote philosopher George Santayana's famous statement, "Those who cannot remember the past are condemned to repeat it." Historians who ascribe to Santayana's philosophy believe that, for example, studying the events that led up to the major world wars or other significant historical events would allow society to chart a different and more favorable course in the future.

Just as difficult as convincing students to realize the importance of studying history is the search for useful and interesting supplementary materials that present historical events in a context that can be easily understood. The volumes in Lucent Books' World History Series attempt to present a broad, balanced, and penetrating view of the march of history. Ancient Egypt's important wars and rulers, for example, are presented against the rich and colorful backdrop of Egyptian religious, social, and cultural developments. The series engages the reader by enhancing historical events with these cultural contexts. For example, in *Ancient Greece*, the text covers the role of women in that society. Slavery is discussed in *The Roman Empire*, as well as how slaves earned their freedom. The numerous and varied aspects of everyday life in these and other societies are explored in each volume of the series. Additionally, the series covers the major political, cultural, and philosophical ideas as the torch of civilization is passed from ancient Mesopotamia and Egypt, through Greece, Rome, Medieval Europe, and other world cultures, to the modern day.

The material in the series is formatted in a thorough, precise, and organized manner. Each volume offers the reader a comprehensive and clearly written overview of an important historical event or period. The topic under discussion is placed in a broad historical context. For example, *The Italian Renaissance* begins with a discussion of the High Middle Ages and the loss of central control that allowed certain Italian cities to develop artistically. The book ends by looking forward to the Reformation and interpreting the societal changes that grew out of the Renaissance. Thus, students are not only involved in an historical era, but also enveloped by the events leading up to that era and the events following it.

One important and unique feature in the World History Series is the primary and secondary source quotations that richly supplement each volume. These quotes are useful in a number of ways. First, they allow students access to sources they would not normally be exposed to because of the difficulty and obscurity of the original source. The quotations range from interesting anecdotes to farsighted cultural perspectives and are drawn from historical witnesses both past and present. Second, the quotes demonstrate how and where historians themselves derive their information on the past as they strive to reach a consensus on historical events. Lastly, all of the quotes are footnoted, familiarizing students with the citation process and allowing them to verify quotes and/or look up the original source if the quote piques their interest.

Finally, the books in the World History Series provide a detailed launching point for further research. Each book contains a bibliography specifically geared toward student research. A second, annotated bibliography introduces students to all the sources the author consulted when compiling the book. A chronology of important dates gives students an overview, at a glance, of the topic covered. Where applicable, a glossary of terms is included.

In short, the series is designed not only to acquaint readers with the basics of history, but also to make them aware that their lives are a part of an ongoing human saga. Perhaps they will then come to the same realization as famed historian Arnold Toynbee. In his monumental work, *A Study of History*, he wrote about becoming aware of history flowing through him in a mighty current, and of his own life "welling like a wave in the flow of this vast tide."

# Important Dates in the History of Twentieth-Century Science

| 1900 | 1910 | 1920 | 1930 | 1940 | 1950 | 1960 | 1970 | 1980 | 1990 |

**1900**
Mendel's experiments showing laws of biological heredity rediscovered.

**1905**
Einstein publishes theory that mass can be changed into energy.

**1907**
Ehrlich formulates first drug to kill microbes inside the human body.

**1913**
Bohr explains the arrangement of electrons in atoms.

**1914**
World War I begins; weapons of war mass-produced for the first time.

**1928**
Fleming discovers penicillin.

**1929**
Hubble discovers that the universe is expanding.

**1938**
Hahn, Meitner, Frisch, and Strassmann discover nuclear fission.

**1943**
Kolff first uses artificial kidney to treat human patient.

**1945**
World War II ends when atomic bombs are dropped on Hiroshima and Nagasaki, Japan.

**1946**
First electronic computer (ENIAC) built.

**1951**
Hopper develops first computer program compiler.

**1953**
Watson and Crick discover structure of DNA; reserpine and chlorpromazine first used to treat mental illness.

**1955**
Mass vaccination against polio begins.

**1957**
*Sputnik I*, first artificial satellite, begins to orbit the earth; Crick describes nature of DNA code.

**1964**
Discovery of background radiation supports Big Bang theory of universe.

**1967**
First successful human heart transplant.

**1969**
Rockets carry human beings to the moon.

**1973**
Scientists combine genes from different kinds of living things.

**1977**
Apple II, a ready-to-use microcomputer, introduced to consumers.

**1982**
Artificial heart implanted in a human.

**1983**
HIV identified as cause of AIDS.

**1989**
Human Genome Project begins.

**1990**
Hubble space telescope launched; first successful human gene therapy begins; World Wide Web first used.

# Introduction

# Science Shapes a Century

About the time the twentieth century began, British statesman Lord Arthur Balfour said of scientists, "They are the people who are changing the world. . . . Politicians are but the fly [device for controlling speed] on the wheel—the men of science are the motive [moving] power."[1]

Balfour spoke even truer than he knew. Since his birth in the mid–nineteenth century, science and technology (the use of science for human benefit) had revealed the cause of deadly diseases, harnessed the power of electricity, and created the first of a dazzling array of synthetic (human-made) chemical compounds. These achievements pale, however, in comparison with what was to come. In the next hundred years, scientists would crack open the fiery hearts of atoms, unravel the chemical threads from which the fabric of life is woven, and hurl human beings into space. British science historian Trevor I. Williams has written, "More has happened in science during this century . . . than in the whole of previous history."[2]

## A Time of Hope and Fear

Many people at the beginning of the century, especially those in the middle and upper classes, thought science and technology were moving society steadily toward a better way of life. On January 1, 1900, the editors of the *New York Times* wrote:

> The year 1899 was a year of wonders . . . in business and production. . . . It would be easy to speak of the twelve months just passed as the banner [best] year were we not already confident that the distinction of highest records must presently pass to the year 1900. . . . The outlook on the threshold of the new year is extremely bright.[3]

Not everyone was so optimistic, of course. Thriving industries brought only misery to the people who worked ten or twelve hours a day in their factories or lost homes or jobs when the factories were built. Some artists and writers, too, viewed the century to come with fear rather than hope. Turn-of-the-century painters such as James Ensor and Edvard Munch filled their canvases with the wavering shapes and distorted faces of nightmare. Contemporary Hungarian author Max Nordau wrote that "the prevalent feeling is of imminent perdition [damnation] and extinction."[4]

Those who sought a reason for their feelings of oncoming doom could point to the rising political tension in Europe. Feelings of nationalism—the belief that one's

## The Greatness of Human Destiny

*At the start of the twentieth century, author H. G. Wells looked ahead with both fear and hope. In this passage, quoted by editor Peter Vansittart in* Voices 1870–1914, *hope won out.*

"One must admit that it is impossible to show why certain things should not utterly destroy and end the entire human race and story; . . . e.g. something from space, or pestilence [epidemic disease], or some great disease of the atmosphere, . . . or some drug or wrecking madness into the minds of men. . . . I have come to believe in other things: in the coherency [order] and purpose in the world, and in the greatness of human destiny. . . . I believe there stirs something within us now that can never die again. . . . A day will come when beings now latent in our thoughts and hidden in our loins [not yet born] will stand upon this earth as one stands upon a footstool, and laugh, and reach out their hands amid the stars."

*H. G. Wells predicted many advancements in science, including atomic energy, computers, and space travel.*

own country, people, and culture are superior to all others—and competition for valuable overseas colonies fostered distrust among the nations of the world, which led them to form complex alliances and begin arming themselves heavily.

Turn-of-the-century predictions of doom seemed to be borne out when nationalist tensions in Europe led to the outbreak of war in 1914. Many historians feel that this war, later called World War I, marked the true beginning of the twentieth century. Like the rest of the new century, the war was shaped by science and technology. War weapons were mass-produced for the first time, for instance, and technology produced new weapons, including poison gas, the tank, and the airplane.

When the war ended in 1918, the factories that had been making war machinery began producing consumer goods instead. Mass-produced, inexpensive automobiles, introduced by Henry Ford in 1908, rolled off reconfigured assembly lines and into the lives of American families in record numbers. Refrigerators, vacuum cleaners,

*Technology boomed with the First World War, when airplanes (left) were developed for warfare. After the war, inexpensive automobiles, introduced by Henry Ford (bottom), began to be mass-produced for consumers.*

and other appliances changed the way women worked in their homes.

Meanwhile, scientists in the 1920s and 1930s changed the way people understood the universe and—often unknowingly—built the foundation for the technology of the future. Astronomers showed that the solar system was a mere dot in a vast star system, or galaxy, which in turn was just one among thousands of galaxies in the universe. Researchers discovered substances capable of killing microorganisms that caused deadly diseases. In Germany, as Adolf Hitler and the National Socialists (Nazis) rose to power and began to prepare for a new war, other scientists learned how to split atoms apart.

During World War II, earlier scientific discoveries were used by all sides to develop technological devices of awesome power. Microbe-killing drugs began to be mass-produced. Rockets, the forerunners of those that would carry humans into space, hurled bombs across Europe. The discovery of a way to split atoms was used to create the biggest bombs of all—the atomic bombs that ended the war and brought a new form of terror into the world.

## Achievements That Changed the World

The rate of scientific and technological achievement seemed to soar in the decade and a half following the end of World War II. So did the public's admiration for science, especially in the United States.

Trevor Williams writes that in the United States during this period,

> Science seemed to promise a universal panacea [cure for all problems]; the scientific bandwagon began to roll and there was a scramble to get aboard.... Politicians of all parties vied [competed] with each other in their promises of support for both scientific research and education.[5]

The people of the 1950s had good reason to admire science and technology. Newly developed antibiotics and vaccines promised an end to microorganism-caused diseases. Drugs had a therapeutic impact on major mental illnesses for the first time. The terrifying power of the atom was harnessed to become a source of cheap energy. Discovery of the "code" used to transmit characteristics from one generation of living things to the next took scientists on the first step toward being able to change that code at will. Admiration of science continued in the 1960s as rockets carried humans into space and, finally, to the moon in 1969.

## Shaping a Century

Advances in the 1970s and 1980s changed society even more. Genetic engineers moved inherited information from one living thing or even one species to another, creating new sources of important medicines and paving the way for improved food crops and animals and even cures for some human diseases. Computers shrank to the size of a desktop and then a large book and became cheap and simple enough for most people to use.

*Science flourished under political and public support following World War II, and the new technology allowed humans to travel to the moon in 1969 (pictured).*

Doctors used new imaging techniques to spot cancers deep in the body or guide their hands during surgical operations.

People of today live in a world that has been formed by modern science and technology. They read or hear news accounts of advances in fields of science, such as quantum mechanics and molecular biology, that did not even exist a hundred years ago. They take medicines or use inventions, such as lasers and computers, that were unknown in their grandparents' time. Many of their jobs, such as software designer or biotechnologist, were created because of scientific advances. In learning about twentieth-century science, young people learn about one of the great forces that shaped this century.

*Chapter*

# 1 Inside the Atom

Early in the twentieth century, renowned British physicist Lord Kelvin wrote, "Physics has come to an end."[6] He thought that the science that had brought him great fame for pioneering discoveries about temperature and heat flow had nothing more to teach the world.

Kelvin could not have been more wrong. Even before his pronouncement, discoveries were taking place that would shatter one of physicists' most basic beliefs, that atoms were the smallest building blocks of matter and could not be divided.

## A New Kind of Ray

In 1895, a German physicist named Wilhelm Röntgen discovered a mysterious new form of radiation that he called X rays. Other physicists hurried to investigate the strange rays. A few months after Röntgen's discovery, French physicist Antoine-Henri Becquerel began studying certain minerals that glowed after being exposed to sunlight. He thought these substances might produce X rays while they glowed. One day he exposed one of these minerals, pitchblende, to sunlight while it lay on a photographic plate that was wrapped in black paper. The paper blocked sunlight, but it would not stop X rays. If the pitchblende was producing X rays, Becquerel reasoned, the rays would darken the film, just as Röntgen had shown that X rays did. Becquerel found that the film did darken, just as he had expected.

*Antoine-Henri Becquerel hypothesized that certain minerals that glowed after being exposed to sunlight were giving off X rays.*

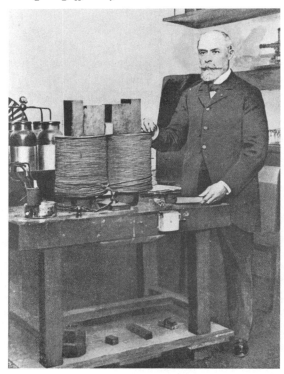

The next day was cloudy, so Becquerel packed away his mineral samples and film. When the sky remained gloomy for several days, though, he decided to open the wrapped film to see if a few faint X rays had penetrated the wrapper even though there had been no sunlight to make the pitchblende glow. He was astonished to find that the film had been strongly darkened. After further investigation, he concluded that the mineral produced rays whether or not it was exposed to sunlight.

Pitchblende contains the element uranium. Becquerel found that the more uranium a pitchblende sample contained, the more intense the radiation was. He concluded, therefore, that the radiation came from the uranium. Moreover, he found that the rays produced by uranium compounds were not the same as X rays. This second kind of new ray soon came to be known as "Becquerel rays."

## The Curies' Work

Two remarkable scientists, Pierre and Marie Curie, became interested in Becquerel rays. French chemist Pierre and his wife, Marie, a physicist born in Poland, were researchers at the Sorbonne in Paris in the late 1890s. The Curies found that compounds of another element, thorium, acted much like uranium compounds in naturally radiating, or giving off, energy—the Becquerel rays. They called this phenomenon radioactivity. The more energy a substance gave off, the more radioactive it was said to be.

One day in 1898, Marie Curie found some samples of pitchblende that were more radioactive than pure uranium. This

*Marie Curie and her husband, Pierre, termed the energy emitted by certain substances radioactivity, and they isolated the radioactive elements polonium and radium.*

suggested that in addition to uranium, pitchblende must contain an unknown, extremely radioactive element. The Curies painstakingly processed tons of pitchblende to extract the new element.

Later in 1898 the couple separated out a black powder that was about four hundred times more radioactive than uranium. It was the element they had been seeking. They named it polonium, in honor of Marie's home country. Even polonium, however, was not radioactive enough to account for the intense energy some of their samples gave off. After more months of work, they isolated an even more radioactive element, which they called radium. For their work in radioactivity, the Curies and Becquerel received the 1903 Nobel Prize in physics.

To the Curies and other physicists who studied the phenomenon, the most puzzling thing about radioactivity was that, according to Marie Curie, the radio-activity of thorium and uranium compounds appears as an *atomic property*. . . . It seems to depend upon the presence of atoms of the two elements in question, and is not influenced by any change in physical state or chemical decomposition.[7]

In other words, because physically, or chemically, changing the substance had no effect on its radioactivity, the radiation must come from the elements' atoms themselves.

This realization led physicists to the astonishing conclusion that atoms, which everyone had assumed were indivisible, must be divisible after all. The atoms of radioactive elements apparently broke apart by themselves, releasing energized streams of fragments. As British physicist Ernest Rutherford exclaimed in amazement to his associate, Frederick Soddy, around 1903:

> Those atoms are exploding! They are bursting open, flinging off bits and pieces of their structure constantly and without let-up. But that's not all! If you ponder the old laws of conservation [of matter], then you have to ask what happens to atoms which throw off these parts of their structure and composition. Yes—you have to admit it—they must change! . . . They are throwing off energy and changing from one element to another.[8]

In ancient times, the forerunners of chemists had searched for a magical "philosopher's stone" that could change one element into another. Now that magic was happening before Rutherford's eyes.

This discovery shattered a second belief of physics and chemistry—the belief that elements could not be transformed—as surely as the radioactive atoms themselves were shattered.

## Mapping the Atom

If atoms could break apart, they must be made of smaller particles. What were those particles?

The first clue came from a device called a Crookes tube, which had helped Röntgen discover X rays. A sealed glass tube from which the air had been removed, a Crookes tube was equipped with a positive and a negative electrode. Physicists noticed that when an electric current was passed through a Crookes tube, a greenish glow surrounded the negatively charged electrode, or cathode. This showed that some kind of radiation was coming from the cathode and traveling to the anode, or positively charged electrode. These so-called cathode rays were different from both X rays and Becquerel rays.

In 1897 a British physicist, Joseph Thomson, showed that cathode rays could be deflected, or sent onto different paths, by an electric charge. A charge would be expected to do this only if the "rays" were actually particles that carried a charge themselves. Because they were deflected very easily, Thomson said, the particles must be much lighter in weight than the only other known charged particles. These other particles were ions, atoms changed by chemical or electrical reactions so that they have a positive or negative electric charge. Normal atoms do not have a charge; they are electrically neutral.

The particles that made up the cathode rays soon became known as electrons.

They proved to be the particles that make up an electric current—the fundamental units of electricity.

Thomson showed that electrons could be produced from a sheet of metal by striking it with ultraviolet light (light with a shorter wavelength than violet, the shortest wavelength of light that the human eye can see). He concluded that the electrons must have come from the atoms of the metal. Therefore, they had to be one of the kinds of particles of which atoms are made. They could not be the only kind, though, because they carried a negative electric charge. Atoms normally have no charge, so an atom had to contain particles with a positive charge that canceled out the negative charge of the electrons.

Thomson suggested that an atom consisted of a positively charged core, or nucleus, with electrons stuck into its surface like raisins in a fruitcake. If one or more electrons were knocked out of the atom, the positive charge of the nucleus would become greater than the negative charges from the remaining electrons. The atom would then become a positively charged ion.

## Quantum Jump

In 1911, Ernest Rutherford revised Thomson's picture of the atom. The atom's nucleus contained most of its mass, Rutherford agreed, but he thought the nucleus was tiny in relation to the size of the atom. He pictured an atom's nucleus as being surrounded by a cloud of even tinier electrons. Electrons' distance from the nucleus, and presumably their weaker "attachment" to it, explained why it was fairly easy to knock electrons out of atoms.

Other scientists' experiments soon suggested that Rutherford, too, was

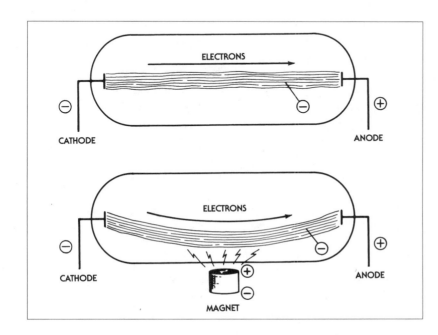

*In a Crookes tube, electricity passes from the cathode to the anode. Joseph Thomson proved that electrons are negatively charged particles by showing how the stream of electrons is attracted to the positive pole of a magnet.*

## The Color of Radium

*In* The Faber Book of Science, *edited by John Carey, Eve Curie tells of a special night in her parents' quest for the secret of radium.*

"The day's work had been hard, and it would have been more reasonable for the couple to rest. But Pierre and Marie were not always reasonable. As soon as they had put on their coats and told Dr. Curie [Pierre's father] of their flight, they were in the street. They went on foot, arm in arm, exchanging few words. After the crowded streets of this queer district, with its factory buildings, wastelands, and poor tenements [slums], they arrived in the Rue Lhomond and crossed the little courtyard. Pierre put the key in the lock. The door squeaked, as it had squeaked thousands of times, and admitted them to their realm, to their dream.

'Don't light the lamps!' Marie said in the darkness. Then she added with a little laugh:

'Do you remember the day when you said to me: "I should like radium to have a beautiful color"?'

The reality was more entrancing than the simple wish of long ago. Radium had something better than 'a beautiful color'; it was spontaneously luminous. And in the sombre shed, where, in the absence of cupboards, the precious particles in their tiny glass receivers were placed on tables or on shelves nailed to the wall, their phosphorescent bluish outlines gleamed, suspended in the night.

'Look . . . Look!' the young woman murmured.

She went forward cautiously, looked for and found a straw-bottomed chair. She sat down in the darkness and silence. Their two faces turned toward the pale glimmering, the mysterious sources of radiation, toward radium—their radium. Her body leaning forward, her head eager, Marie took up again the attitude which had been hers an hour earlier at the bedside of her sleeping child."

wrong. Instead of moving in a random cloud, electrons apparently circled the nucleus as planets orbit the sun. Furthermore, they were grouped in several specific orbits, or shells, each of which could hold only a certain number of electrons.

This new picture of electrons in shells explained some basic facts of chemistry.

Chemists had known since the mid–nineteenth century that certain elements had features in common. Some of these common features proved to be due to similarities in the outermost electron shell of the elements' atoms. Elements that almost never combined with other elements to form compounds, for instance, turned out to have an outer shell that was full (contained the maximum number of electrons allowable for that shell). Elements with just one electron in their outer shell, on the other hand, tended to give up that electron to atoms of other elements. In doing so, they easily formed compounds.

Why did electrons move only in certain orbits? The answer turned out to lie in an observation that a German physicist named Max Planck made in 1900. Planck studied the way a black object that had been heated radiated energy in the form of light back into space. He found that dif-

*Niels Bohr thought that electron orbits could be explained by using quantum theory, or the idea that energy could be absorbed or sent out only in fixed amounts.*

*In the early 1900s, scientists theorized that an atom, shown here, was made up of different shells of electrons orbiting a nucleus.*

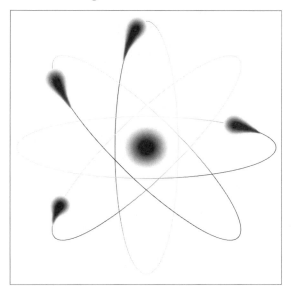

ferent wavelengths of light, such as red and violet, had different amounts of energy. Furthermore, energy could be absorbed or sent out only in fixed amounts that could be thought of as separate little packets. He called each packet a quantum, from a Latin word meaning "How much?"

In 1913, Danish physicist Niels Bohr used Planck's quantum theory to explain electrons' orbits. Bohr said that electrons could gain or lose energy only in quantum-sized packages, and these changes in energy corresponded to changes in orbit. If an electron loses a quantum's worth of energy, it jumps inward by one shell. On the other hand, if an electron absorbs a quantum of energy, it jumps one shell outward.

## Into the Atom's Core

By using radiation to bombard atoms, Ernest Rutherford was able to identify positively charged particles he called protons, from a Greek word meaning "first." Protons were much more massive than the negatively charged electrons, but they had the same amount of electrical charge. They were the predicted particles that balanced the electrons' negative charge, making atoms electrically neutral.

Each element had a different number of protons in the nucleus of its atoms. The number of protons became known as the element's atomic number. It normally equaled the number of electrons in orbit around the nucleus. One mystery remained, however: Experiments showed that the nuclei of all atoms except hydrogen were heavier, or more massive, than could be accounted for by "weighing" the protons alone.

During the 1930s, physicists created "atom smasher" machines to study atomic particles further. These machines used magnetic fields to speed up fragments of atoms given off by radioactive elements and hurl them into other atoms. Researchers using these machines found that when atomic fragments called alpha particles were slammed into a piece of the element beryllium, a new kind of very penetrating radiation was produced. The radiation was so powerful that it could knock protons out of the carbon atoms in paraffin, a kind of wax. Magnetic fields did not affect it, so the radiation could not consist of charged particles (electrons or protons). Magnetic fields, like electric fields, affect only charged particles.

In 1932, British physicist James Chadwick claimed that the radiation was a stream of a new kind of atomic particle. Because the particles could knock protons out of other atoms, they had to be as massive as protons. They were electrically neutral, so Chadwick called them neutrons. Physicists concluded that the cores of all atoms except hydrogen contained both protons and neutrons. Together, protons and neutrons made up most of an atom's mass. The combined number of protons and neutrons in the nucleus of an element's atoms equaled that element's atomic weight.

*Ernest Rutherford discovered that the nucleus of an atom contained positively charged particles called protons.*

## Splitting Atoms

In the early 1930s, Enrico Fermi in Italy and Otto Hahn in Germany, among others,

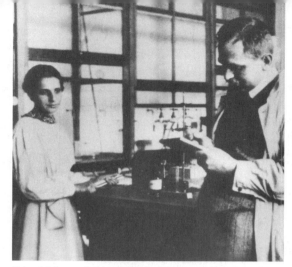

*Lise Meitner (left) and Otto Hahn (right) found that bombarding uranium atoms with neutrons split the atoms' nuclei in half and produced energy.*

began bombarding atoms of uranium, the heaviest natural element, with neutrons. They hoped the atoms would absorb neutrons and thereby change into atoms of new, artificial elements that were heavier than uranium. But the resulting atoms did not seem to have atomic weights near that of uranium. Indeed, the atoms in some of Hahn's experiments with Fritz Strassmann had atomic weights that were only about half that of uranium.

Hahn described his results to an old friend and colleague, Austrian physicist Lise Meitner, in December 1938. Meitner, a Jew, had fled to Sweden to escape Nazi persecution in Germany. She discussed Hahn's experiments with her nephew and fellow physicist, Otto Frisch. The two came to the astonishing conclusion that Hahn's bombardment was splitting the nuclei of uranium atoms in two, turning about 1 percent of each atom's mass to energy in the process. Frisch named the splitting process fission, borrowing the biological term for the multiplication of cells by splitting.

No one had dreamed such a thing might occur—but someone *had* predicted what would happen if it could be done. In 1905 a former Swiss patent office clerk named Albert Einstein had theorized that mass could be turned into energy. He described the relationship between the two in a famous equation, $E = mc^2$ (energy equals mass times the square of the speed of light). Because the speed of light is a very large number, Einstein's equation meant that a tiny mass could be converted into a huge amount of energy. Meitner and Frisch calculated that each splitting uranium atom yielded as much energy as twenty million times its weight in the powerful explosive TNT.

Further, physicists soon realized that if amounts of uranium greater than a certain critical mass (well above the amount Hahn and Strassmann had used) were bombarded, the uranium might be arranged so that when one atom broke apart, the fragments would include neutrons that could in turn split more atoms, making more neutrons that would split still more atoms. The resulting chain reaction might be used to generate tremendous amounts of energy. This energy could provide a steady supply of electric power—or a source of devastating destruction.

## "Brighter than a Thousand Suns"

In 1939, as Europe went to war, Albert Einstein wrote a letter to U.S. president Franklin D. Roosevelt warning that

> it may be possible to set up a nuclear chain reaction by which vast amounts

*Albert Einstein predicted the results of what became known as nuclear fission; he also warned that the technology could be used for mass destruction.*

of power and large quantities of radium-like elements might be generated . . . and it is conceivable—though much less certain—that extremely powerful bombs of a new type might be constructed.[9]

During World War II, the United States set up a successful crash program to develop such bombs. On July 16, 1945, when an atomic bomb was first tested, Richard Feynman, one of the project scientists present at the test, wrote to his mother, "We jumped up and down, we screamed, we ran around slapping each other on the backs."[10] But the blazing flash of the bomb, "brighter than a thousand suns," made Robert Oppenheimer, director of the project, think of a description of the terrible goddess Kali in the Hindu scriptures of India: "Now I am become death, the destroyer of worlds."[11]

Less than a month later, the United States dropped atomic bombs on the cities of Hiroshima and Nagasaki, forcing Japan, its last remaining enemy in World War II, to surrender. The bombs ended the war sooner than it might otherwise have ended. They also ended the lives of more than one hundred thousand people.

## Nuclear Engineering

The image of the shattered atom was everywhere in the 1950s and 1960s, symbolizing both good and evil. In 1954 the U.S. Navy launched *Nautilus*, the first submarine to be powered by a reactor that converted the energy from controlled fission of atomic nuclei into steam to run turbines. Nuclear power was a clear advantage for ships because it allowed them to sail for many months without refueling. Soon both the United States and the Soviet Union were building fleets of nuclear-powered submarines, some of which carried missiles with nuclear warheads.

Nuclear energy also lent itself to peaceful applications. The world's first commercial nuclear power plant went into service at Shippingport, Pennsylvania, in 1957.

Naval and commercial nuclear power reactors in the United States use water for cooling and heat transfer. Heat from the fission reaction in the reactor core heats water in a closed tube, or loop. This heat in turn heats water in a different tube, creating steam that spins a turbine. Two separate tubes must be used in the reactor because the water in the first tube becomes radioactive after passing through

> **The Energy of the Future**
>
> *In* From Creation to Chaos: Classic Writings in Science, *Bernard Dixon quotes British statesman Winston Churchill's 1932 prediction about the atomic-powered world of the future.*
>
> "We know enough to be sure that the scientific achievements of the next fifty years will be far greater, more rapid, and more surprising than those we have already experienced. . . . Nuclear energy is incomparably greater than the molecular energy we use today. The coal a man can get in a day can easily do five hundred times as much work as the man himself. Nuclear energy is at least one million times more powerful still. If the hydrogen atoms in a pound of water could be prevailed upon to combine together and form helium, they would suffice to drive a thousand-horse-power engine for a whole year."

the reactor core, and it must be kept out of the surrounding environment. (Exposure to radioactive material can cause cancer, birth defects, and other health problems.) A loss of water from this tube would also cause dangerous overheating of the reactor core, potentially leading to an explosion that would release large amounts of radioactive material.

Reactor designers try to ensure that if the water from the primary loop does leak out (perhaps because a valve gets stuck or

*Devastation in Japan from an atomic bomb blast in World War II. The United States used the development of the nuclear chain reaction to create a powerful weapon, bomb Hiroshima and Nagasaki, and end the war.*

a pipe ruptures), the number of neutrons in the reactor core will go down, quickly stopping the fission reaction.

Nuclear accidents can result from both design and human errors. A reactor built by the Soviet Union at Chernobyl lacked any "fail-safe" design features and melted down in 1986, causing the worst nuclear accident in history. An American reactor at Three Mile Island had a serious accident in 1979 because operators mistakenly turned the cooling system off.

Designers have responded to safety concerns about nuclear reactors in two general ways. They have tried to simplify reactor design by, for example, reducing the number of valves and other controls used. The fewer parts a machine has, the fewer ways it can fail. They have also tried to design safety systems that rely on natural laws rather than human decisions to make the reactor "do the right thing." Emergency cooling water can be stored so that gravity rather than a pump will make it flow when needed, for instance.

Critics argue that even if so-called inherently safe nuclear power plants can be built, the problem of what to do with their waste remains. As a reactor operates, some of the material around its core becomes radioactive. Also, not all the nuclear fuel is consumed in the reaction. This leftover "spent fuel" is still very radioactive. Spent fuel and other radioactive waste have to be stored until the radiation decays to a safe level—which for some materials means many thousands of years. Meanwhile, they must not be exposed to the air or seep into the groundwater.

Because of these concerns (and the relatively low cost of competing fuels, such as coal), the nuclear power industry has not expanded much in recent years. Certain environmental problems may lead people to give nuclear power a second look, however. Nuclear power stations do not produce chemical pollution or add to possible global warming as coal-powered plants do, for instance.

*The 1986 nuclear reactor meltdown at Chernobyl, shown here after the accident, demonstrated the catastrophe of nuclear reactor failure and the consequent release of radioactive substances into the environment.*

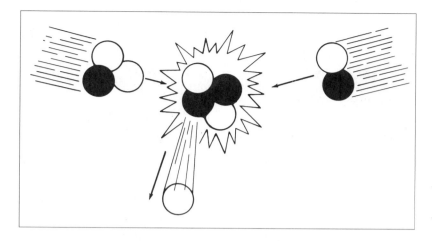

*In this depiction of the fusion of two types of hydrogen nuclei, the white spheres represent neutrons and the black spheres represent protons. One neutron is expelled with great force, producing the energy in the reaction.*

Some scientists believe that a safer form of reactor might work by nuclear fusion rather than fission. In a nuclear fusion reaction, nuclei of two atoms are forced together to form a third kind of atom. Fusion reactions release even more energy than fission reactions, but they are harder to start and control. A nuclear fusion reactor would use a relatively cheap and safe fuel (probably a form of hydrogen) and would produce waste that is less radioactive. A number of different designs for fusion plants have been explored, but none appears practical for use now or in the near future. How diligently these and other approaches to safer nuclear power are pursued in the next century may depend on how serious shortages of conventional fuels, such as oil and coal, or environmental problems become.

## The Atomic Zoo

While some physicists have focused on harnessing atomic power for human use, others have continued to explore the atom for its own sake. By 1960 physicists had found so many kinds of particles in what some have called the "atomic zoo" that they almost despaired of finding a simple set of underlying principles that would allow them to understand the atom.

Starting in 1961, however, American physicist Murray Gell-Mann suggested that the force that holds the atomic nucleus together results from combinations of particles he called quarks. Surprisingly, quarks have fractional electric charges—one-third or two-thirds that of an electron. They have other characteristics that nuclear physicists describe with such whimsical terms as *bottom, top, up, down, truth,* and *beauty.*

Many physicists think that quarks may at last be the truly fundamental, unbreakable particles of matter. Quarks cannot be observed directly, however. Theories describing them and other subatomic particles continue to grow more complex, suggesting that much remains to be discovered. Perhaps in the twenty-first century the true picture of the atom will at last emerge.

# Chapter 2
# The Expanding Universe

Hold a finger a few inches in front of your face. Close one eye and look at the finger. Notice where your finger appears against the background of objects on the other side of the room.

Now repeat the process with the other eye. Your finger will appear to have moved against the background. This is because your eyes are several inches apart, and switching from one eye to the other changes the angle at which you see your finger. This difference in angle is called parallax.

Once someone has measured the angle to an object from two points in space, a type of mathematics called trigonometry can be used to calculate the object's distance. The ancient astronomer Ptolemy used this method to show that the moon was about 240,000 miles away from the earth. After telescopes were invented in the sixteenth century, astronomers could measure the much smaller parallaxes of the sun and planets. They used these to determine these bodies' distances from the earth.

## How Far the Stars?

In 1838, a German astronomer named Friedrich Wilhelm Bessel, using an extremely accurate telescope and many observations, succeeded in measuring the parallax of a star other than the sun for the first time. The difference in angle turned out to be equal to the diameter of a quarter held by someone ten miles away! Bessel calculated that the star, 61 Cygni, was an incredible sixty-four trillion miles away, or about nine thousand times the width of the solar system. Suddenly the universe seemed a much bigger place.

Gradually, astronomers measured the parallax of other nearby stars. Because trillions of miles proved an inconvenient unit of measurement, they began to use the light-year as a unit of distance instead. A light-year is the distance that light, moving at about 186,000 miles a second, travels in one year—about 5.9 trillion miles. Proxima Centauri, the star that is our sun's nearest neighbor, is 4.3 light-years away.

Because the parallax of even nearby stars is so small, another yardstick was needed to measure the distance of farther stars. Astronomers found such a yardstick in certain stars that grew dimmer and then brighter in a predictable cycle. They named these stars cepheid variables after Delta Cephei, the first such star they identified. In 1912, American astronomer Henrietta Leavitt studied the smaller of two glowing patches of stars called the

*This time-lapse exposure of the Milky Way includes the trail of a communications satellite. By studying the distances of stars from Earth, astronomers identified the Milky Way as well as other galaxies.*

Magellanic Clouds. She measured twenty-five cepheids in the cloud and found that the brighter a cepheid seemed to be, the longer was its period, or the time it took to go from dim to bright to dim again.

Once this relationship between brightness and period in cepheids was established, astronomers used it to estimate the distance of cepheids from the solar system. If two cepheids had the same period, they had the same absolute magnitude, or true brightness. Because apparent brightness decreases with the square of distance, if one of two cepheids with the same period appeared to be four times dimmer than the other, it had to be twice as far away as the brighter star.

By studying the cepheids in spherical clouds of stars called globular clusters, Harlow Shapley, an American astronomer, was able to draw a scale map of the stars relatively near the solar system. His map showed that the solar system is far out along one arm of a vast, spiral-shaped group of stars called a galaxy. Part of the disk-shaped center of this galaxy appears in the sky as the Milky Way, and astronomers give the whole galaxy that name.

While some astronomers continued mapping the Milky Way, others began to look at small fuzzy patches of light called nebulae (from the Greek word for clouds). One especially bright patch, the Andromeda Nebula, caused a debate among astronomers. Was it a nearby gas cloud or something much larger and more distant?

In 1924, Edwin Hubble, perhaps the greatest of American astronomers, looked at Andromeda through a new one-hundred-inch telescope at Mt. Wilson, California. Studying photos taken through the telescope's powerful lens, he could see individual stars—including some cepheids—shining at the edges of the nebula. Using the cepheids as a yardstick, he and other researchers eventually determined that the distance to Andromeda was about 2.5 million light-years. If it was that far away, Andromeda had to be another galaxy.

Since the 1930s, astronomers have discovered thousands of galaxies of different sizes and shapes. Some exist in clusters called supergalaxies. Today the size of the known universe has been extended to about ten billion light-years.

*Andromeda, shown here with two companion galaxies, is the nearest spiral-shaped galaxy to the Milky Way. Discovered by Edwin Hubble in 1924, it was thought to be about 2.5 million light-years away.*

## What Are Stars Made Of?

Astronomers had discovered that the universe had billions of stars, but they did not know what stars were made of or what made them shine. Before these questions could be answered, new scientific instruments and a new kind of physics would have to be invented.

In the nineteenth century, astronomers began using a device called a spectrograph to break the light from stars into a spectrum, a series of bands and lines like the colors in a rainbow. By analyzing the spectra (plural of *spectrum*) of different stars, they could determine which chemical substances were glowing on each star's surface. It turned out that the most common element in stars was hydrogen, followed by a newly discovered element called helium.

Spectra could also be used to determine the surface temperature of stars. Annie J. Cannon and other researchers grouped stars into classes according to their surface temperature. In order of decreasing temperature the spectral classes are O, B, A, F, G, K, M, R, N, and S. (Astronomy students often learn this order by memorizing the sentence "Oh, be a fine girl [or guy]—kiss me right now, sweetheart!") The sun belongs to class G, about midway between the hottest and coolest stars.

In 1911 and 1913, Danish astrophysicist Ejnar Hertzsprung and American Henry Russell independently discovered a relationship between the brightness and the temperature of stars. The Hertzsprung-Russell diagram, a chart based on their research, plots stars according to their surface temperature (decreasing from left to right) and their absolute brightness or magnitude (increasing from bottom to top). Along a diagonal line from upper left to lower right called the main sequence, stars could be arranged from hot, bright, blue ones to cool, dim, red ones. Generally, large stars are hotter and brighter, while small stars are dimmer and cooler. However, astronomers also discovered red giants, which are large and bright but cool, and white dwarfs, which are small and dim but relatively hot.

## The Life Cycle of Stars

Astronomers could now track the color, temperature, and composition of stars, but they could not answer a more basic question: What kept the stars shining? If the sun were something like a huge lump of burning coal, it would have burned itself out after only a few thousand years, yet obviously it had been shining for far longer than that.

*An astronomer uses a spectrograph, a device that breaks the light from stars into a spectrum, which can then be analyzed to determine the chemical content of the stars and their surface temperatures.*

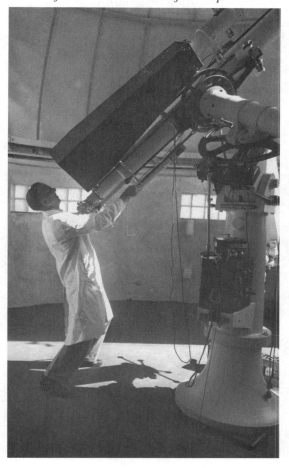

Hermann von Helmholtz, a German physicist, suggested that the sun produces energy as its mass gradually falls toward its center, just as a rock gains energy as it falls to earth. Lord Kelvin estimated that this process might allow the sun to shine for about fifty million years. If that were true, the earth, having cooled after the solar system was formed, could be only about twenty million years old. However, geologists (scientists who study the earth) had already shown that the planet was at least a billion years old.

For a long time, physicists were unable to explain how the sun could have stayed warm so long. Then, in the 1930s, they discovered nuclear reactions and the tremendous energy that such reactions could produce from even small amounts of matter. If nuclear reactions were taking place on stars, that could explain how stars remained so hot for such a long period.

In 1938, physicist Hans Bethe showed that under the immense pressures and temperatures found in the cores of the sun and other stars, hydrogen nuclei could fuse together to form helium nuclei. Since its spectrum had revealed that the sun was mostly hydrogen, it had enough fuel to carry out such fusion reactions for billions of years.

The fuel supply of the stars is not endless, however. During the 1950s, British astrophysicist Fred Hoyle developed a new understanding of how stars change during their lifetimes. He and several coworkers found that, almost like living things, stars are born, mature, grow old, and eventually die. The product of their nuclear reactions changes at each stage.

According to Hoyle's theory, about five billion years from now the sun will start running out of hydrogen. Its helium core will grow and the sun will cool and expand, becoming a red giant. Eventually it will grow large enough to swallow up the inner planets, including the earth.

As the sun's core becomes still more massive, the sun will begin to contract, or shrink, and heat up. At that stage a new kind of nuclear reaction will create carbon by fusing helium atoms. The cycle will repeat as the sun first expands and cools, then shrinks and heats up again. Each time it will form elements heavier than those that existed before. Eventually the nuclear reactions will no longer perpetuate themselves, though, and the sun will contract to a tiny, dense, relatively cool body about the size of our earth—a white dwarf star. The white dwarf will continue to glow faintly for many billions of years.

The sun will have a quiet death, but stars that start out much larger than the sun can run their nuclear cycle all the way up to reactions that produce iron. When the outer part of a star with an iron core at a temperature of billions of degrees starts to collapse, it fuses all at once, creating an immense nuclear explosion called a supernova. This explosion shatters most of the star, leaving only a tiny remnant surrounded by a cloud of glowing gas.

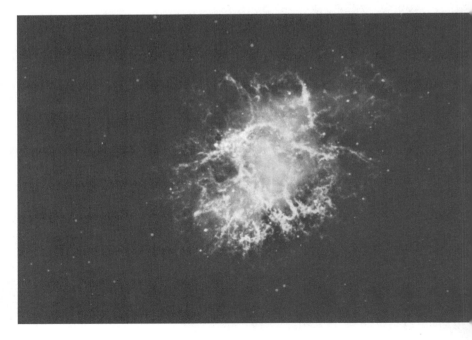

*Astrophysicist Fred Hoyle (top) theorized that stars have a life cycle that includes creation, maturity, and death. When a star's life cycle ends with a supernova, only a small remnant and a cloud of gas (bottom) remain.*

# Long, Cold Nights

*In* Lonely Hearts of the Cosmos, *science writer Dennis Overbye describes the painstaking way astronomers had to work in the early twentieth century, before the movements of telescopes could be controlled automatically by computer.*

"To make sharp photographs the telescope had to track the stars as they moved across the sky. Unfortunately, you couldn't trust even the finest telescope to track the stars faithfully, let alone an antique like this [60-inch telescope at Mt. Wilson]. As the telescope heeled over, it would flex [bend slightly], minute imperfections in its gears either speeding it ahead or dropping it behind the stars it was following; also differential refraction [bending of light] would shift the apparent location of the stars in the sky. So you had to stand on the moving platform in the sky with your eye to the separately moving eyepiece, with a control paddle in your hands hitting buttons to goose the telescope [speed up its movement] or slow it down.

Guiding could be literally painful. On cold nights tears would freeze one to the eyepiece. When the telescope moved, the [observatory] dome had to move, to keep its slit opening in front of the telescope, and the observer's platform had to move as well. Over the course of a night these movements, all independent of each other, could gradually scrunch an astronomer's neck and back into celestial pain."

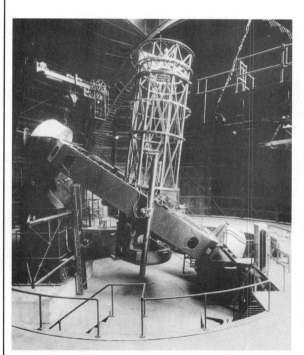

*A view of the interior of Mt. Wilson Observatory. Manually controlled telescopes were not as accurate or as easy to operate as later telescopes, which are controlled by computers.*

Science broadcaster Nigel Calder, in his radio program *The Key to the Universe*, pointed out:

> In a sense human flesh is made of stardust. Every atom in the human body, excluding only the primordial [original] hydrogen atoms, was fashioned in stars that formed, grew old, and exploded most violently before the Sun and the Earth came into being. The explosions scattered the heavier elements as a fine dust through space. By the time it made the sun, the primordial gas of the Milky Way was sufficiently enriched with heavier elements for rocky planets like the earth to form. And from the rocks atoms escaped for eventual incorporation in living things.[12]

## The Expanding Universe

Stars, it seemed, could grow, change, and even die. Did the universe as a whole also change? Did stars and galaxies move around in space? At first these questions seemed impossible to answer. The motion of a few nearby stars could be measured, but most stars (let alone other galaxies) were much too far away for any possible motion to be seen.

One clue could be used to reveal motion, however. In 1842 an Austrian physicist named Christian Doppler explained why the whistle on a train approaching an observer first seems to rise in pitch, and then, as the train passes the observer, to fall in pitch. This "Doppler shift" is caused by sound waves from the whistle bunching together as they reach the observer from the approaching train and then stretching out as the locomotive moves away.

*After astronomers discovered that galaxies were moving away from Earth, Edwin Hubble devised a means of determining the rate at which they moved away.*

Light also exists as waves of radiation, and waves of light from an approaching object (such as a star) bunch together just as sound waves do. This makes the star appear more violet than it really is. When a star is moving away from an observer, on the other hand, it appears more red. The faster the star is moving, the greater the shift in wave frequency and thus in color.

Astronomers began measuring Doppler shifts in galaxies around 1912. They quickly came to a startling conclusion: With a few exceptions, such as Andromeda, the galaxies all appeared to be moving away from Earth. In 1929, Edwin Hubble announced that he had found a relationship between how far a galaxy was from Earth and how fast it was moving away. If one galaxy was twice as far away as another galaxy, he said, it moved away at twice the speed. It was as though the universe were a huge balloon with galaxies painted on it like dots. The balloon kept being blown larger and larger, and all the galaxies therefore moved farther and farther away from each other. In short, Hubble said, the universe was expanding.

In the early 1960s, well-known astronomers offered two opposing theories to explain why the universe was expanding. Fred Hoyle's "steady state" theory claimed that new matter was appearing between the galaxies at a rate too slow to detect, pushing the galaxies farther and farther apart. A rival theory, popularized by George Gamow, maintained that the universe had originated in a stupendous explosion, which came to be known as the "Big Bang." The expansion of the universe, Gamow said, is the continuing result of that explosion.

At first there seemed to be no way to tell if either theory was right. Since the 1950s, however, scientists had been building special telescopes to detect the radio waves that are constantly being given off by stars and other objects in space. Gamow predicted that if the Big Bang had really happened, a background level of radio waves should have been left behind. In 1964, radio astronomers detected this radiation. Most astronomers consider this discovery to be conclusive evidence that the Big Bang theory is correct.

## Black Holes and the End of the Universe

In the 1930s Indian-American astrophysicist Subrahmanyan Chandrasekhar had discovered that if a star had more than 1.44 times the mass of our sun, it could not turn directly into a white dwarf as it aged. The star first had to lose enough mass to fall below this limit. If it could not do this, the star would keep collapsing until it became a neutron star—a star whose atoms are stripped of their electrons and jammed together to create an extremely dense form of matter.

Some researchers speculated that some stars might become even more dense than neutron stars. If a star became dense enough, its gravity would be so great that light could no longer escape its surface. The star would become a totally dark object—a "black hole." Inside the

*An artist's depiction of the Big Bang. The Big Bang theory states that the universe was created in one gigantic explosion. The propulsion of matter away from the explosion still continues, resulting in an expanding universe.*

black hole would be a condition called a singularity, in which matter, space, and time were no longer separate and the laws of physics did not seem to apply.

During the 1960s, British astrophysicist Stephen Hawking carried on research into black holes despite suffering from a crippling condition called ALS, or Lou Gehrig's disease. One day, while riding home on a train, Hawking had an idea. Suppose the superdense point that exploded to create the universe in the Big Bang was also a singularity like that in a black hole?

To show how the Big Bang singularity might have shaped the universe, Hawking combined Einstein's theories of relativity and ideas from quantum mechanics to draw a kind of map that showed how time and space would change as the universe grows. The map is like a globe of the earth, with the Big Bang at the north pole and the end of the universe—its possible collapse or "big crunch"—at the south pole. In between, the universe expands until it reaches its largest extent (like the equator on the globe) and then gradually contracts.

But is this what will really happen? That depends on whether the universe has enough total mass to cause the force of gravity to eventually overcome the force

---

### An Amazing Universe

*Physicist Stephen Hawking begins his book* A Brief History of Time *by reminding us how extraordinary our world truly is.*

"We go about our daily lives understanding almost nothing of the world. We give little thought to the machinery that generates the sunlight that makes life possible, to the gravity that glues us to an Earth that would otherwise send us spinning off into space, or to the atoms of which we are made and on whose stability we fundamentally depend. Except for children (who don't know enough not to ask the important questions), few of us spend much time wondering why nature is the way it is; where the cosmos [universe] came from, or whether it was always here; if time will one day flow backward and effects precede causes; or whether there are ultimate limits to what humans can know. There are even children, and I have met some of them, who want to know what a black hole looks like; what is the smallest piece of matter; why we remember the past and not the future; how it is, if there was chaos early, that there is, apparently, order today; and why there *is* a universe."

from the Big Bang and start pulling the universe back together again.

Astronomers have tried to estimate how much mass the universe contains. Even counting mass that is too dark to see but that can be detected by its gravitational effects, there seems to be only about a tenth of the mass needed to halt the universe's expansion. Many astronomers believe that additional mass exists in forms that cannot yet be detected, however. The search for the "missing mass" of the universe still goes on.

## Exploring the Solar System

Compared to other scientists, astronomers suffered from a handicap. The objects they studied—stars, planets, and other heavenly bodies—were too far away to touch and sample. They could only observe the light and other radiation that reached the earth from these bodies. The earth's atmosphere blocks some wavelengths of light, however, and makes images in even the best earthbound telescopes waver.

On October 4, 1957, this situation began to change. *Sputnik I,* the first artificial earth satellite, was boosted into space by a Soviet rocket and began to circle the earth. This satellite carried little in the way of scientific instruments, but its successors allowed astronomers to begin exploring space. The American *Explorer I* satellite, for example, discovered bands of highly charged particles that surrounded the earth. They were named the Van Allen belts after their discoverer.

People feared at first that the Van Allen belts would make the space environment too dangerous for human exploration, but fortunately, that did not prove to be true. Beginning in 1961, humans followed their satellites into space. People walked on the moon for the first time in 1969.

Meanwhile, robot space probes pushed farther and farther into the solar system. For instance, the American probe *Mariner 2* visited cloud-shrouded Venus in 1962, followed by a succession of Soviet and American probes. Science fiction writers had often pictured Venus as a hot but Earth-like world, covered by jungles, swamps, or perhaps a planetwide sea. Probe data, however, revealed that Earth's sister planet was bone dry and had a temperature of 750° F.—hotter than a broiling oven.

Mars had inspired the human imagination since American astronomer Percival Lowell had claimed to see a network of irrigation canals crisscrossing the red planet's surface around the start of the twentieth century. After several probes had flown by to take photographs or crashed into the Martian surface, the National Aeronautics and Space Administration (NASA) began the Viking project, an ambitious program to explore Mars in greater detail. In 1976, while two Viking spacecraft orbited Mars and made a photo survey, their landing modules touched down on different parts of the planet's surface. The lander cameras sent back to Earth striking pictures of a rock-strewn land with a red sky overhead.

The Viking landers also contained miniature biological laboratories. Mechanical arms took soil samples and exposed them to carbon dioxide and other nutrients to see whether plants or other life forms in the soil would begin to grow. After some confusion over observed evidence of growth, however, scientists concluded that the results of these experiments could be

*A two-meter-wide boulder dominates this view of the surface of Mars. A recent discovery has prompted speculation that some form of life may have once existed on Mars.*

explained by chemical processes that did not involve life.

In 1996, however, researchers announced an astonishing finding. About fifteen million years ago, an asteroid collided with Mars and sent pieces of Martian rock hurling into space. About thirteen thousand years ago one of these rocks fell to Earth in Antarctica. Inside the rock are chemical traces and tiny gaps similar to those left by the bacteria that were the earliest form of life on Earth. This interpretation is controversial, but it has again raised the possibility that some form of life exists (or once existed) on Mars. Plans for a new mission to Mars are now being discussed, including the possibility of a lander that could collect samples, return them to an orbiting spacecraft, and bring them to Earth for study.

The Pioneer, Voyager, and Galileo probes visited the outer planets in the 1970s and 1980s. These probes showed Jupiter's moons to have a striking diversity of surface features, including active volcanoes in one case. Today every planet in the solar system except Pluto has been photographed and mapped. Astronomy has become a "hands-on" science.

Space astronomy has moved far beyond the solar system. With the launch of the Hubble space telescope in 1990, astronomers gained an instrument that escaped the interference of the earth's atmosphere. After visiting astronauts fixed a faulty mirror in 1993, the telescope's power to magnify details was comparable to being able to read the license plate on a car thirty miles away.

## Mission to Earth

One common criticism of the space program has been that the money spent on exploration could be better spent for more earthly needs. Space, however, has proven to be an excellent vantage point from which to study our home planet.

Since the 1960s, sensitive instruments in satellites have been used to examine nearly every aspect of the earth. Infrared (heat-sensing) instruments have tracked the activity of volcanoes and the movement of ocean currents. Weather satellite pictures are now part of every TV news broadcast. Special cameras, often adapted from those used in military spy satellites, have followed the movement of animal herds in Africa, measured the shrinkage of the rain forest in Brazil, and discovered the outlines of ancient cities buried beneath desert sands. Thus the exploration of the solar system from space has included Earth, too.

Measurements of the movements of Earth's crust performed by satellites have also helped to support a new understanding of Earth that geologists gained earlier in the century. Earth scientists have known since the nineteenth century that the details of the planet's crust have altered over millions of years. In the early 1960s, however, they started to accept the more startling proposal that the continents themselves have moved during geologic time.

German meteorologist (a scientist who studies weather and climate) Alfred Wegener first proposed this idea early in the

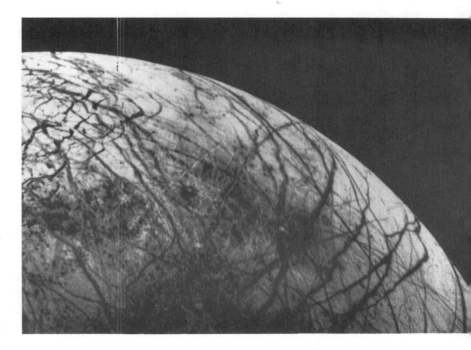

*A 1979 photo of one of Jupiter's moons, Europa. The dark lines snaking across the surface are thought to be cracks in the icy crust that have been filled in with materials from the planet's interior.*

century. Wegener pointed out, for example, that the west coast of Africa and the east coast of South America looked as if they would fit together like pieces of a jigsaw puzzle. This, he said, was because the two continents had once been united. Wegener claimed that two hundred million years ago, all of Earth's continents had been part of a single landmass.

For several decades, geologists laughed at Wegener's idea of "continental drift." They could think of no force that would cause the continents to move in the way he described. After World War II, however, improved instruments made possible better mapping of the ocean floor and more accurate geologic dating. The new and better information inspired a

## A Living Planet

*Scientist and science writer Lewis Thomas, in* The Lives of a Cell, *describes the beauty of the earth as seen from space. Some other scientists have come to share Thomas's conviction that the earth is "alive."*

"Viewed from the distance of the moon, the astonishing thing about the earth, catching the breath, is that it is alive. The photographs show the dry, pounded surface of the moon in the foreground, dead as an old bone. Aloft, floating free beneath the moist, gleaming membrane of bright blue sky, is the rising earth, the only exuberant thing in this part of the cosmos. If you could look long enough, you would see the swirling of the great drifts of white cloud, covering and uncovering the half-hidden masses of land. If you had been looking a very long, geologic time, you could have seen the continents themselves in motion, drifting apart on their crustal plates, held aloft by the fire beneath. It has the organized, self-contained look of a live creature, full of information, marvelously skilled in handling the sun."

*Earth's swirling clouds and constantly changing, active surface give it the image of a living entity.*

*Alfred Wegener, a meteorologist in the early 1900s, believed that the continents had once been part of a giant landmass. His idea was the starting point for modern plate tectonics.*

theory called plate tectonics, which revived and added to Wegener's proposals.

The plate tectonics theory states that the earth's crust is divided into a number of plates. The plates float on a layer of fiery liquid rock called the mantle, slowly moved by currents in the mantle just as ocean currents move floating matter in the sea. The parts of the plates that rise above the ocean's surface form continents and islands.

Along the edges of the plates, both on land and under the sea, the crust cracks open. At spots where plates are drifting apart, molten rock from the mantle rises up into the cracks. The rock cools and hardens, forming new crust. Old crust, in turn, is forced down into the mantle at spots where plates are pushed together. Movement at the edges of plates often causes earthquakes and volcanic eruptions.

The acceptance of plate tectonics has been just one of the ways in which scientists' understanding of the earth has changed during this century. Twentieth-century science has brought a new realization that the earth, like the solar system and the universe of which it is a part, is an ever-changing place.

# Chapter 3
# The Code of Life

On back-to-school night, it is often easy to tell which children belong with which parents. Family members frequently look alike. They may even behave in similar ways. People have always known this, but until the twentieth century, they had little idea how or why these resemblances occurred. The way living things inherit and are shaped by information from their parents is the subject of genetics, a science that did not exist until the twentieth century.

## Discoveries in a Garden

The story of genetics really began in a nineteenth-century monastery garden. In the town of Brno, now in the Czech Republic but at the time part of the empire of Austria-Hungary, a monk named Gregor Mendel tended pea plants year after year. Some of Mendel's plants were tall, others short; some produced smooth peas, some wrinkled ones. Mendel carefully mated plants with different characteristics, or traits, and observed the traits of the offspring they produced.

Mendel found that the seven traits he studied were passed from parents to offspring in predictable ways. A plant, he decided, inherits a "factor" from each parent that determines the form of a particular trait. In the case of height, for instance, a plant can inherit either two tallness factors, two shortness factors, or one factor of each kind. If the plant inherits two factors of the same kind, it will match its parents.

*Austrian monk Gregor Mendel tended and studied pea plants in the nineteenth century and decided that factors that determined certain traits were passed from parents to offspring.*

# Results of Mendel's Pea Plant Experiment

Mendel devised a diagram to help others visualize the inheritance of particular characteristics in offspring. In his diagram, he assigned letters to each characteristic and used a capital letter to designate a dominant trait and a lowercase letter to designate a recessive trait. A capital *A* identified the dominant trait for tallness in pea plants, for example. A lowercase *a* identified the recessive trait for shortness. Here is how he showed the results of crossing a tall plant (AA) with a short plant (aa), and crossing a tall plant (Aa) with another tall plant (Aa):

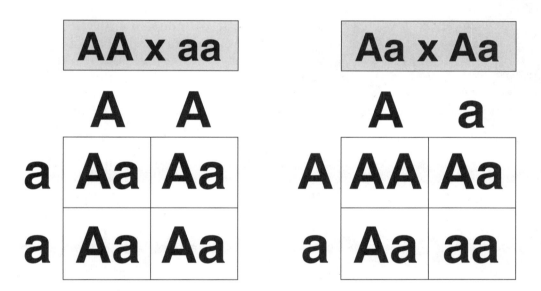

If it inherits one factor of each kind, it will be tall because, as Mendel said, the factor for tallness is dominant, or rules, over the factor for shortness. The shortness factor is recessive. It "draws back" and does not show itself if a tallness factor is present.

The shortness factor does not disappear, however. It can be passed on to offspring. Mendel observed that when two plants, each with one factor for tallness and one for shortness, are mated, one out of every four of their offspring will be short because those offspring will have inherited a factor for shortness from both parents.

Mendel published a paper describing these and other results of his pea

*Charles Darwin thought that the environment dictates which traits are kept as living things adapt and survive.*

experiments in a small scientific journal in 1866. At the time, no one paid much attention to this account or recognized its importance. By the turn of the century, though, scientists were increasingly interested in understanding how traits were inherited, in large part because British biologist Charles Darwin's theory of evolution, controversial when it was first published in 1859, had become widely accepted. To fully understand evolution, scientists needed to understand heredity.

Darwin said that nature, over long periods of time, selected the traits that living things would have, just as farmers and ranchers mated domestic plants and animals to develop improved forms that had the most desirable characteristics of both parents. If the climate in a certain area became drier, for example, living things in that area that could survive with little water would be more likely to live long enough to have offspring than those that required a lot of water. Over time, therefore, the number of living things in the area that had inherited traits that helped them do without water would grow, while the number that had not inherited such traits would shrink. No one, including Darwin, however, understood the mechanism of heredity that drove this process of "natural selection."

In 1900, three scientists—one in the Netherlands, one in Germany, and one in Austria—independently did experiments similar to Mendel's. They discovered the same rules he had worked out. When they checked to see what earlier scientists had written about plant breeding and inheritance, each unearthed Mendel's long-forgotten paper. The three cited Mendel's work when they described their own. All reported that their findings simply duplicated those of the obscure monk, who by then had been dead for sixteen years.

## Chromosomes and Genes

Soon after Mendel's work was rediscovered, several scientists suggested that his factors might be connected with chromosomes, threadlike bodies in the nucleus, or center part, of cells. Scientists had learned about sixty years earlier that cells are the microscopic units of which the bodies of living things are made.

Chromosomes exist in pairs. German scientists had discovered in the late nineteenth century that just before a cell divides, its chromosomes split down their length. The two halves of each chromosome pull apart. The cell then divides, producing two "daughter" cells. Each daughter cell gains a set of chromosomes just like the set in the parent cell. This meant that the chromosomes had to duplicate themselves somehow when they split.

Something different happens in sex cells, the egg and the sperm that can unite to become a new living thing. When these cells form, the pairs of chromosomes separate, but the individual chromosomes do not split or duplicate. As a result, each sex cell gets only half as many chromosomes as a body cell. When an egg and a sperm

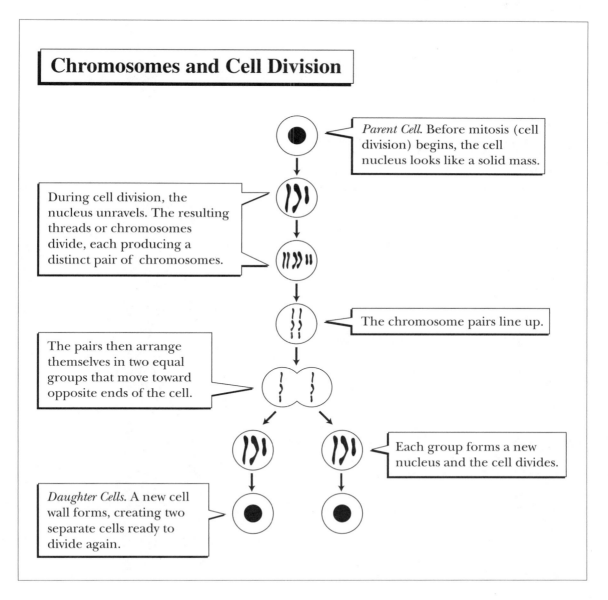

**Chromosomes and Cell Division**

*Parent Cell.* Before mitosis (cell division) begins, the cell nucleus looks like a solid mass.

During cell division, the nucleus unravels. The resulting threads or chromosomes divide, each producing a distinct pair of chromosomes.

The chromosome pairs line up.

The pairs then arrange themselves in two equal groups that move toward opposite ends of the cell.

Each group forms a new nucleus and the cell divides.

*Daughter Cells.* A new cell wall forms, creating two separate cells ready to divide again.

unite, the resulting fertilized egg again has the full number of chromosomes. Half—one chromosome in each pair—come from the mother and half from the father.

In the way they were passed on to offspring, chromosomes behaved just like Mendel's factors. Scientists quickly saw, though, that the chromosomes themselves could not be the factors. Different kinds of living things have different numbers of chromosomes (humans have forty-six, for instance), but the numbers are always small, a few dozen at most. It was obvious that most living things have far more than a few dozen traits. Each chromosome, therefore, had to contain many factors. After 1909, scientists began calling these units genes, from a Latin word meaning "I produce offspring." But no one had any idea what a gene looked like or how it worked.

## Race to Unravel a Molecule

Chromosomes contain two kinds of complex chemicals: proteins and nucleic acids. Most important chemicals in cells had proved to be proteins, and until the early 1940s, most geneticists (scientists who studied biological inheritance) thought that genes would be proteins, too. Experiments with bacteria, however, turned their attention to the other possible candidate, nucleic acids.

Scientists began to suspect that the secret of the way nucleic acids carried information from parents to offspring lay in the structure of the molecules, or combinations of atoms, that make up these substances. In particular, they began looking at the structure of the best-known nucleic acid: deoxyribonucleic acid, or DNA. They knew that each large, complex molecule of DNA was made up of several kinds of smaller molecules. The DNA molecule was long and thin, with a "backbone" of molecules called phosphates and sugars. Some biochemists suspected that this backbone had the shape of a coil, or helix. Four kinds of molecules called bases—adenine, guanine, cytosine, and thymine—were attached to the backbone. No one knew, though, exactly how the bases were attached or how any part of this structure worked in cells.

In the early 1950s, several groups of scientists tried to determine the structure of DNA, with an urgency some viewed as a race on an international scale. The winners of the race proved to be twenty-five-year-old genetic researcher James Watson, an American, and thirty-seven-year-old Francis Crick, an English geneticist, colleagues at England's Cambridge University.

*James Watson (left) and Francis Crick (right) were the first scientists to determine the structure of the DNA molecule.*

## Discovering the Secret of Life

*In* The Double Helix, *James Watson recalls the day he and Francis Crick solved the structure of DNA—the "secret of life."*

"When I got to our . . . office the following morning, I quickly cleared away the papers from my desk top so that I would have a large, flat surface on which to form [models of] pairs of bases held together by hydrogen bonds. Though I initially went back to my like-with-like prejudices [the idea that two bases of the same kind paired with each other in the DNA structure], I saw all too well that they led nowhere. . . . [I] began shifting the bases in and out of various other pairing possibilities. Suddenly I became aware that an adenine-thymine pair . . . was identical in shape to a guanine-cytosine pair. . . .

Upon his arrival Francis [Crick] did not get more than halfway through the door before I let loose that the answer to everything was in our hands. . . . A few minutes later he spotted the fact that . . . both [kinds of] pairs could be flipflopped over and still have their . . . bonds facing in the same direction. This . . . strongly suggested that the backbones of the two chains must run in opposite directions. . . .

At lunch Francis winged into the Eagle [a local public house or bar] to tell everyone within hearing distance that we had found the secret of life."

*James Watson displays his model of the DNA molecule.*

After working with metal models and studying photographs of DNA molecules made with a special X-ray technique, Watson and Crick decided that DNA's "backbone" formed not one helix but two. The bases, they concluded, lay in pairs between the two intertwined backbones, like steps on a spiral staircase. Guanine always paired with cytosine, and adenine always paired with thymine. The two spiral backbones coiled in opposite directions, so the order of base-pair "steps" was the same regardless of which end of the DNA molecule one started with.

Watson and Crick described their groundbreaking theory of DNA's structure in a one-page scientific paper published on April 25, 1953. "It has not escaped our notice," they wrote, that the structure they proposed "suggests a possible copying mechanism for genetic material."[13] When chromosomes are ready to split, the two scientists explained in a second paper published a few weeks later, the weak chemical bonds holding the base pairs together break down. As a result, each long DNA molecule splits down the middle like a zipper unzipping. Each base on both halves of the molecule then attracts its "mate" from among free-floating materials in the cell nucleus. A new backbone also forms. The finished product is two DNA molecules, both identical to the original one. One of the two molecules ends up in the set of chromosomes going to one daughter cell, and the other molecule goes to the other cell.

## Genetic Blueprints

Watson and Crick had solved a central mystery of genetics: how the material in chromosomes duplicates itself. Of equal importance, however, was the question of how DNA molecules carry information. Scientists in the 1930s had found evidence that genes determine traits by providing blueprints for proteins. Geneticists came to define a gene as the instructions for making one protein. It was not clear, however, what form those instructions took.

Like DNA, proteins are large, complex molecules composed of smaller molecules. The smaller molecules in proteins are called amino acids. The problem was that DNA contains only four kinds of bases, while proteins are made up of twenty kinds of amino acid. A single DNA base, therefore, could not stand for an amino acid.

In 1957 Francis Crick pointed out that sets of three bases would produce more than enough possible combinations to represent all the amino acids. "It is like a code," Crick explained in a letter to his son. "If you are given one set of letters you can write down the others."[14] Each set of three bases, Crick said, made up one "letter" in DNA's code, just as several dots and dashes in a certain order stand for a single letter in Morse code. During the 1960s, Crick and other scientists did experiments to learn which sets of bases stood for which amino acids. They found also that, rather than standing for amino acids, some base combinations signaled the beginning or end of a gene. Each DNA molecule contains thousands of genes.

At about the same time, Crick and others worked out the process by which a gene produces a protein. The cell first uses the DNA of the gene as a model, or template, to create a strand of a second nucleic acid, RNA (ribonucleic acid). RNA is similar to DNA except that it has a different kind of sugar in its backbone and a fifth base, uracil, is substituted for the thymine in DNA. The RNA strand contains the same sequence of bases as the DNA that made it.

DNA cannot leave the cell nucleus, but RNA can. The newly made strand of RNA moves into the cytoplasm, the jellylike substance that makes up the body of the cell. Many chemicals, including all the types of amino acid, float free in the cytoplasm. Certain cell chemicals "recognize" the amino acids specified by the RNA and bring them, a molecule at a time, to the RNA molecule. Other chemicals string the amino acid molecules together, in the order specified by the RNA, to form the protein.

Watson and Crick's discovery and the advances that followed it revolutionized genetics. Scientists in genetics and a new field of study called molecular biology, which focused on physical and chemical processes in cells, made great strides in the next decades. They learned, for instance, how to find out what kind of protein a particular gene made and, conversely, how to locate and work out the base sequence of the gene that made a particular protein. In 1967 a leading biologist, looking back on this time,

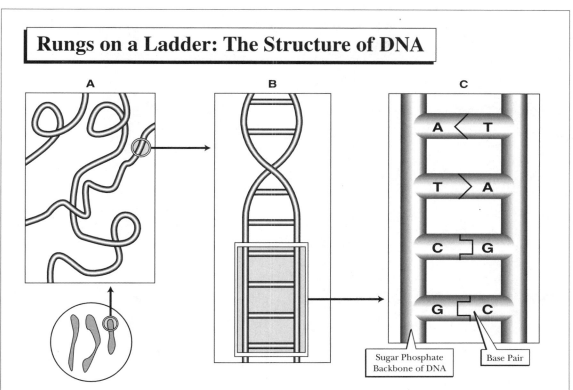

**Rungs on a Ladder: The Structure of DNA**

A. A chromosome is a chainlike strand of DNA, which contains many genes.

B. When the chromosome is greatly magnified under a microscope, it looks like a long ladder that is twisted into a double helix. The twisting allows these amazingly long strands to fit inside a single tiny cell.

C. The sides of the DNA ladder are made of sugar and phosphate molecules. Between the two sides are rungs made up of the four base pairs—AT, TA, GC, and CG. The letters stand for the four bases that make up the pairs: adenine, guanine, cytosine, and thymine. A single strand of DNA may contain billions of rungs. The different arrangements of these four base pairs are codes that call for different combinations of amino acids. Amino acids combine to make up proteins, which, in turn, combine to form the endless variety of features that make up every living thing. Each sequence of base pairs that contains the instructions for making a single protein is called a gene.

said, "more progress has been made in our understanding of the fundamental . . . mechanisms of life in the last 20 years than in all the previous history of biology."[15]

## Genetic Clocks

Geneticists and molecular biologists applied their new knowledge in a variety of ways. For instance, they learned more about evolution by comparing DNA from related groups of animals and humans. Geneticists interested in the origins of human beings found that only 1.6 percent of the genes in the genome (complete collection of genes) of modern humans differ from the genes of chimpanzees. By estimating the rate at which mutations, or random changes in genes, occur, it is possible to calculate the amount of time required to accumulate the number of mutations that make humans different from chimps. At the beginning of that time, humans and chimpanzees would have had an ancestor in common. There are a number of uncertainties in such calculations, but they point to human ancestors' separating from the ancestors of chimpanzees and other apes six to eight million years ago.

Other DNA studies have helped scientists decide when modern humans came into existence. Most DNA research focuses on the DNA in the nuclei of cells, but scientists have learned that mitochondria, organelles (small bodies) in the cytoplasm, also contain DNA. This mitochondrial DNA, or mtDNA, mutates about ten times faster than nuclei DNA. It therefore can be used as a "genetic clock" to determine the timing of events that occurred relatively recently in evolutionary history, such as the times when different populations of humans split off from one another.

Comparing the mtDNA of modern people from around the world, researchers have found that Africans have the most genetic variation, reflecting the most mutations. This suggests that Africans have had the most time to accumulate changes and are thus the oldest population of humans. These studies of genetic changes suggest that modern humans appeared in Africa about two hundred thousand years ago.

Unlike nuclear DNA, mtDNA is passed directly from mother to child, without any mixing with the father's DNA. Because mtDNA is passed through the female line, the theory based on its measurement has been called the "African Eve" theory. Some scientists support the theory, but others have challenged it. Controversy over its validity continues today.

## Genes and Disease

As researchers became able to identify individual genes and work out the sequence of their bases, they learned more about how genes can make people sick. Damage by sunlight, certain chemicals, or other factors may cause a mutation in a gene, which is a change in the sequence of its bases. Such a genetic change is like a misspelling in a word. Cells have mechanisms for repairing mutations, just as misspellings are usually corrected before a book is printed. Sometimes, though, a mutation is not corrected. Then the cell either cannot make the protein for which the gene was a blueprint or makes an altered, defective form of the protein. A change in even one base can sometimes cause this to happen. Lacking the healthy form of the protein, the cell may malfunction or even die.

If the mutated gene becomes part of the genes that are passed on to offspring, the offspring may not be able to make the gene's protein either. Lack of the protein may result in an inherited disease. Inherited diseases affect a little under 1 percent of babies born. They include such serious and sometimes fatal conditions as sickle-cell anemia, a blood disease, and cystic fibrosis, a disease that affects the lungs and makes breathing difficult. Geneticists have been able to identify many of the genes whose defects cause particular inherited diseases. Researchers have also learned that genes can play a role in diseases that are not directly inherited. Some genes can make it more likely that a person will develop heart trouble, for instance, though they do not actually cause the illness.

One of the most startling discoveries about genes and disease was the finding that damaged genes are the basic cause of cancer. Oncogenes, or genes capable of causing cancer, were found first in viruses that cause cancer in animals. Then in 1975, Michael Bishop and Harold Varmus of the University of California at San Francisco found harmless genes similar to these oncogenes in healthy cells. "We carry the seeds of our cancer within us,"[16] Bishop has said. At some point in the cancer-causing viruses' evolutionary past, he believes, the viruses picked up these genes from cells they infected.

Later, scientists learned that the harmless or normal forms of oncogenes produce cell growth. Other genes shut them off at a certain stage of development. If the normal genes mutate in certain ways, however, they can no longer be shut off, and the uncontrolled cell growth of cancer results. Cancer can also arise when the "shut-off" genes, sometimes called anti-oncogenes, are damaged or destroyed. Scientists now believe that factors in the environment, such as chemicals in food or cigarette smoke, produce the gene changes that cause most cancers. These changes occur in body cells during a person's life rather than being inherited. However, people can inherit genes that make them likely to develop some types of cancer.

## Mapping the Human Genome

In an effort to better understand genes and their role in disease, in 1989 the United States and a number of other countries launched a project to map the entire human genome—some one hundred thousand genes, or three billion bases, in all. The project, called the Human Genome Project, will locate each human gene on its chromosome and determine the sequence of bases in it. The Human Genome Project is estimated to cost $3 billion and is scheduled to be completed in the year 2003. Such an ambitious undertaking has been made possible by advances in technology that allow most steps in the identification and sequencing of genes (determining the sequence of their base pairs) to be done automatically, by machine. Today, science writer Richard Golob says, "The techniques of gene isolation [identification] and cloning [copying, part of the sequencing procedure] . . .

*The Human Genome Project, aimed at locating each human gene on its chromosome and determining the sequence of bases in it, may provide a map of health for a person's life.*

have been refined to the point that they can easily be performed by a student in a high school laboratory."[17]

Identification of the complete collection of human genes will allow doctors of the future to analyze each individual's genome in detail. People will be identified by their genetic profiles even more accurately than by fingerprinting, because each person's collection of genes (unless he or she has an identical twin) is unique. (All humans have the same types of genes, but most genes occur in several forms, so individual genomes differ.) A genetic profile will offer a "health map" for a person's life. Leroy Hood, the inventor of a machine that automatically determines the sequence of bases in a gene, says that once the human genome is mapped,

> When a baby is born, we'll "read out" his genetic code, and there'll be a book of things he'll have to watch for. This has the potential to do enormous good. If you have a propensity [tendency] toward heart disease . . . or cancer, you could modify your diet or change the environmental substances you're exposed to [to decrease your risk of getting the disease].[18]

## Genetic Engineering

While some scientists have been identifying and mapping genes, others have been finding ways to change them. In 1973 Paul Berg and Stanley Cohen at Stanford University and Herbert Boyer at the University of California at San Francisco learned how to move genes from one kind of living thing to another. They used natural chemicals called restriction enzymes to snip stretches of DNA from viruses or bacteria. They then used other chemicals called ligases to insert the pieces into the genomes of other viruses or bacteria. Each gene made its trademark protein in its new location. For instance, when a gene whose code specified a protein that destroyed a certain antibiotic was put into bacteria that had not been resistant to that antibiotic, the bacteria became resistant. Later, scientists learned how to insert genes not only into bacteria but into higher organisms, including human beings. Today they can combine genes from very different living things.

## Drug Factories

These "gene-splicing" techniques produced a new field of technology called genetic engineering or biotechnology. Genetic engineering's first benefit to people came from putting human genes into bacteria. The chosen genes made body chemicals that were useful in medicine, such as the hormone insulin, which is used to treat diabetes. Bacteria multiply very quickly, and some kinds can be grown in huge numbers in vats. Genetically engineered bacteria became "factories" that produced these vital chemicals at a cost much lower than that of extracting the substances from the bodies of animals or people. The genetically engineered substances were also safer because they could not be contaminated by dangerous viruses sometimes found in blood or tissue.

Today, plants and even farm animals have been turned into drug factories by inserting the genes of other species, even human genes, into their cells. "Pharming," as the process is sometimes called (a blend of *farming* and *pharmaceuticals*, meaning medicines or drugs), has created goats that produce a chemical that can save lives after heart attacks, pigs that make a key part of a human blood substitute, and tobacco plants that make a chemical needed in sunscreen, to name just a few.

Scientists are also using genetic engineering to improve domestic animals and crop plants themselves. For instance, genes that make plants resistant to certain diseases have been inserted into some crop plants. Other plants have been given genes from bacteria that make a chemical that kills pest insects. (The chemical has no effect on humans or other mammals.) Science writer Nigel Calder says that genetic alteration of plants is expected to be "by far the biggest money-spinner" among genetic engineering applications in the near future, adding a predicted $20 billion yearly to the value of crops worldwide by the year 2000.[19] Farm animals are being genetically altered, too. Genes that result in the production of leaner meat, for instance, have been engineered into cattle, pigs, and other animals raised for human consumption.

## Gene Therapy

The greatest hope for genetic engineering is that it will be able to replace defective genes that cause disease. This has already been done in animals and in a very small number of human beings. Such treatment is called gene therapy.

One of the first people to undergo gene therapy for an inherited disease was a little girl named Ashanthi DeSilva. Because of a single defective gene, she was born without the ability to make a chemical called adenosine deaminase, or ADA. Lacking this substance, her body's immune system was totally unable to fight off disease. Other children born with this rare defect had had to spend their lives inside plastic "bubbles" to protect them from microbes in the air. Even then, such children seldom lived to adulthood.

Beginning in 1990, when Ashanthi was just four years old, W. French Anderson and R. Michael Blaese of the National Institutes of Health removed some white blood cells from Ashanthi's body and grew them in laboratory dishes. The researchers introduced a harmless virus, which had been engineered to contain the human gene that produced ADA, into the cells. The cells were then reinjected into Ashanthi's blood. There they began manufacturing ADA, allowing the girl's immune system to work normally. Because even healthy blood cells do not survive long, the treatment at first had to be repeated every few months. But now, Ashanthi needs a treatment only once a year. She goes to school and leads a normal life.

This treatment has been used successfully in a few other children with Ashan-

### An Age of Intervention

*In "The Recombinant-DNA Debate," an article that appears in* Genetics: Readings from *Scientific American, Clifford Grobstein addresses both supporters and critics of genetic engineering.*

"Society has entered an age of intervention, in which the automatic operation of natural processes is increasingly . . . brought consciously into the orbit of human purpose. . . .

The policy challenge we face . . . is whether we can create institutions able to transform the fruits of an age of reason into the achievements of an age of intervention. There are voices today urging us not only to eschew [give up] conscious intervention but also to distrust and limit the uses and consequences of reason itself. Perhaps it needs to be restated that it was, after all, natural selection that evoked . . . human knowledge and judgment. . . .

The concept and control of the double helix signal a new frontier of biocultural progression. A . . . vision that includes both 'creative pessimism' and 'creative optimism' is now required. Neither alone can do justice to the profound revelations human beings have recently experienced. A single eye is particularly limited in yielding depth and perspective. For the age of intervention at least two are needed."

*A comparison of a genetically altered cotton leaf (right) with a leaf that has been damaged by insects (left). The leaf on the right has been genetically manipulated to resist insects.*

thi's disease, and scientists hope to treat other inherited diseases similarly. So far, however, this very complex procedure can be used to treat only diseases caused by a defect in a single, known gene. (Many diseases result from defects in several genes or in genes that have not yet been identified.) Widespread use of genetic engineering to treat human disease is probably decades away. Nevertheless, it is one of the great hopes that geneticists have for the twenty-first century.

# Chapter 4
# Magic Bullets

By the end of the nineteenth century, doctors had learned the causes of many diseases that spread from person to person. Such diseases were caused by microbes—bacteria, viruses, and other living things too small to see without a microscope. Doctors knew how to kill microbes on objects such as surgical instruments and bandages, but once disease-causing microbes invaded a person's body, little could be done to stop them.

All that changed in the early twentieth century, when the first drugs that could kill microbes inside the body were developed. The first person to find such a drug was a German medical researcher named Paul Ehrlich. In 1907 Ehrlich formulated an arsenic compound that killed the microbes responsible for syphilis, a deadly sexually transmitted disease. A refined version of the drug, which was marketed as Salvarsan, was first used on human beings in 1909. Ehrlich called it a "magic bullet" because it aimed at and destroyed a particular type of microbe.

In the 1930s, German scientists developed drug "bullets" that could hit a wider range of targets. The sulfa drugs, made from a compound called sulfanilamide, first appeared in 1932 after Gerhard Domagk, a chemist working for a German dye company, found that a red dye called Prontosil stopped the growth of several kinds of common bacteria. The active ingredient in the dye proved to be sulfanilamide. The sulfa drugs were the first of a class of man-made compounds in wide use today in treating bacterial infections.

*Paul Ehrlich developed an arsenic compound in 1907 that killed the microbes responsible for syphilis. Prior to his discovery, doctors did not know how to kill disease-causing microbes once they entered the body.*

## An Accidental Discovery

Even before Domagk began working with Prontosil, a British scientist had an accident that would lead to an even better weapon against microbes. In 1928, Alexander Fleming was growing colonies of bacteria in dishes in his laboratory at St. Mary's Hospital in London. When he came back from a three-week vacation, he found a patch of bluish mold growing on one of the dishes.

Fleming started to throw the "spoiled" dish away . . . and then stopped. Most of the dish's surface looked cloudy because of the bacteria growing on it. Around the patch of mold, though, Fleming saw a clear ring. That must mean, he realized, that no bacteria were growing near the mold. The mold must make some chemical that stopped them.

When Fleming put the dish under a microscope, he could see the bacteria dying. He wrote later:

> When I saw those bacteria fading away, I had no suspicion that I had got a clue to the most powerful therapeutic [healing] substance yet used to defeat bacterial infections in the human body. The appearance of the culture [bacterial colony] was such, though, that I knew it should not be neglected.[20]

Fleming's mold was a common type that grew on stale bread and rotting oranges. Its scientific name was *Penicillium notatum*. Fleming therefore called its microbe-killing excretion penicillin. He found that penicillin stopped the growth of a variety of bacteria in his laboratory dishes. He doubted that it would be useful in medicine, though, because each mold

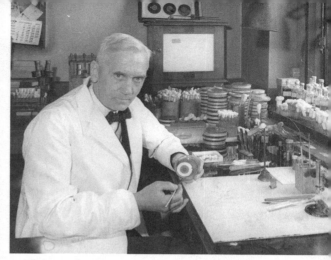

*In 1928 Alexander Fleming accidentally discovered that a chemical produced by mold killed bacteria. He called the substance penicillin.*

made only a tiny amount of the substance. Fleming therefore published an account of his discovery in a scientific journal in 1929 and then more or less forgot about it.

## Penicillin Goes to War

No one else paid much attention to Fleming's find, either, until the late 1930s. At that time Howard Florey, an Australian-born researcher, and biochemist Ernst Chain, a Jewish refugee from Nazi Germany, were at England's Oxford University studying antibiosis, the process by which some microorganisms kill others. Looking for examples of antibiosis in the scientific literature, they unearthed Fleming's report, much as turn-of-the-century biologists had rediscovered Gregor Mendel's research on heredity.

Like Fleming, Florey and Chain did not sense the importance of the discovery of penicillin at first. They knew, though, that streptococcus, the chief kind of

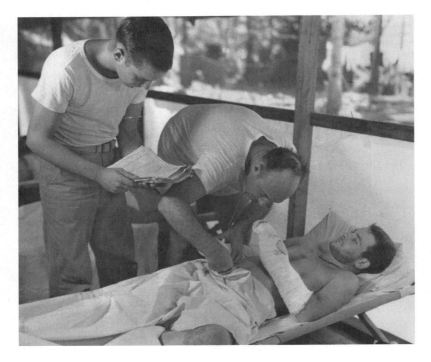

*A soldier receives penicillin at a Marine Corps facility in 1944. The first widespread use of penicillin was during World War II, when the drug was administered to sick soldiers.*

bacteria that penicillin killed, was a common cause of wound infection. With war in Europe looming, they could not help wondering whether the compound might save wounded soldiers' lives. They began testing it in animals. In May 1940 they showed that penicillin could stop fatal streptococcus infections in mice. "It looks like a miracle!"[21] Florey exclaimed. The first tests on humans worked equally well.

By that time, Britain was under attack by Germany. It had no resources to spare for research and development of a new drug, even a miracle drug. In July 1941, therefore, Florey flew to the United States to meet with government officials and scientists from drug companies. After the United States entered World War II in December 1941, American scientists began a crash program to develop ways to mass-produce the *Penicillium* mold in large vats. They also found ways to extract penicillin from the mold efficiently. By 1943, enough penicillin was being made to treat half a million soldiers a month.

Scientists soon discovered a host of other antibiotics, or drugs from chemicals made by one kind of microbe to destroy another. These compounds worked so well that most Americans and Europeans in the 1950s and 1960s thought the war against microbe-caused disease was all but won.

## Defending Against a Crippler

Killing invading microbes is only one way to protect the body against them. Another is to strengthen the body's defenses so that it can destroy the invaders on its own. This is the purpose of vaccines.

A vaccine introduces killed or weakened disease microbes into the body. Exposure to these harmless forms teaches the immune system, the body's natural defense against disease, to recognize this type of microbe. If full-strength forms of the same microbe enter the body later, the immune system is ready to destroy them. Vaccines against several important human and animal diseases were developed in the nineteenth and early twentieth centuries.

By the middle of the twentieth century, the disease against which Western doctors most eagerly sought a vaccine was polio. This contagious disease is caused by a virus, so antibiotics could not cure it. It killed or crippled thousands of people, mostly children, each year. In 1955 there were fifty-eight thousand new cases of polio in the United States alone. The disease's victims included Franklin D. Roosevelt, president of the United States from 1933 to 1945. Roosevelt contracted polio in 1921, when he was thirty-nine years old, and was never afterward able to walk without support.

In 1938 President Roosevelt founded a private organization, the National Foundation for Infantile Paralysis (another name for polio), or NFIP. This organization controlled most polio research in the United States. It soon became better known by

## A Miracle Cure

*In* The Health Century, *Edward Shorter describes one of the first successful uses of penicillin in the United States. Shorter quotes Samuel Mines's book* Pfizer: An Informal History.

"By 1943 word of penicillin is getting out. People know it can save dying relatives and call desperately for it. Yet the government has reserved the [small] supply for research on civilians and for military needs. What, therefore, was Pfizer [an American drug company that had just started making penicillin] executive John Smith supposed to do when Dr. Leo Loewe, a physician at the Brooklyn Jewish Hospital, came to him and pleaded for penicillin to 'save the life of a doctor's small daughter' dying of an infection of the valves of the heart? . . . Smith said he'd come and see the little girl. His heart melted [and he gave Loewe the penicillin].

'For three days Dr. Loewe administered penicillin to the dying child . . . , dripping it into a vein from a hanging bottle twenty-four hours a day. Her condition improved.' They continued giving her enormous amounts. 'During those days, as the color came back to her face, Smith . . . came, day after day, to watch a miracle—the first human being to be snatched from death's sure grip by his company's own penicillin.'"

the title of its most successful fund-raising campaign, the March of Dimes.

The first problem researchers faced was figuring out how to grow the huge amounts of polio virus that would be needed to mass-produce a vaccine. Viruses reproduce only inside cells. At first the researchers could get polio virus to grow only in the spinal cords of living monkeys. In the late 1940s, however, John Enders, Thomas Weller, and Frederick Robbins found a way to make the virus grow in skin and muscle cells in test tubes.

## Salk vs. Sabin

Different scientists had different ideas about the best kind of polio vaccine to use. Jonas Salk of the University of Pittsburgh began work on a vaccine made from killed polio virus. Albert Sabin of the University of Cincinnati, on the other hand, thought a vaccine made from weakened living virus would be better. At first there was no scientific reason to favor one approach over the other. However, New York attorney Basil O'Connor, head of the NFIP, liked Salk better than he liked Sabin. Salk therefore won the backing of the NFIP, with all its money and publicity, and his vaccine was finished and tested first.

On April 12, 1955, the NFIP's evaluating committee announced that Salk's vaccine appeared to be both safe and effective. Factory whistles and church bells rang out across America to herald the news. Mass vaccination then began. By May 7, four million doses of the Salk vaccine had been distributed.

The Salk vaccine was not perfect. It had to be administered by a series of injections and was not completely effective

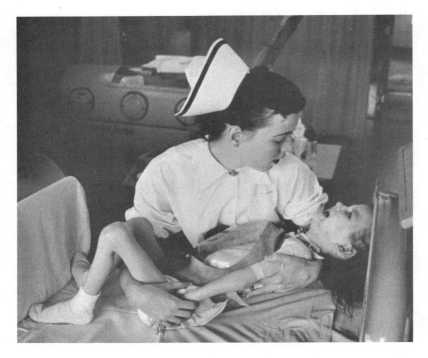

*The newly developed antibiotics of the mid–twentieth century were powerless against the ravages of polio, a contagious, crippling, and sometimes deadly disease that struck mainly children.*

*The Salk vaccine for polio, given in a series of injections (left), was distributed in the United States in 1955. It was eventually replaced by the Sabin polio vaccine, which could be taken orally and was effective in a single dose (right).*

against one major type of polio virus. By contrast, Albert Sabin's vaccine, which had been fully developed by the time the Salk vaccine was put into use, could be taken by mouth and was effective in a single dose. It worked against all three major types of polio viruses. In spite of these advantages, the NFIP refused to support mass testing of the Sabin vaccine in the United States. Countries ranging from Sweden to the Soviet Union, however, tried the vaccine with good results in the late 1950s. The Sabin vaccine finally began to be used in the United States in the early 1960s. It eventually replaced the Salk vaccine for many purposes. Thanks to these two vaccines, there have been no new natural cases of polio in the United States since 1979.

As antibiotics and vaccines reduced deaths from microbe-caused diseases in industrialized countries, more people in those countries began living into old age. Medical researchers naturally turned their attention to illnesses that usually strike older people, such as heart disease and cancer. Drugs did not cure these disorders as successfully as they cured diseases caused by microbes. Researchers did, however, find drugs that slowed, controlled, or sometimes even stopped these serious illnesses.

The first drug used successfully against a variety of cancers was related to a weapon of chemical warfare. The weapon was a poison gas called mustard gas, which had been used in World War I. During World War II, three Yale University scientists were studying mustard gas and a chemical relative, nitrogen mustard, to learn more about their effects. They happened to give nitrogen mustard to a mouse with advanced lymphoma, a kind of cancer in which certain cells in the immune system multiply uncontrollably and form a tumor (mass of cells).

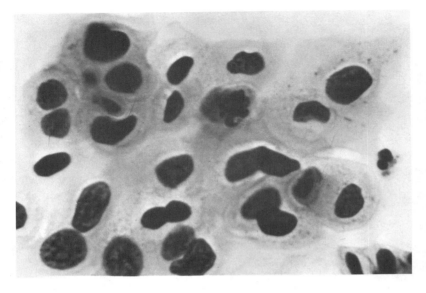

*Nitrogen mustard was found to decrease the size of masses of cancerous cells like these. This discovery, made during World War II, led to other anticancer drugs, including one effective against a type of childhood leukemia.*

"The tumor completely regressed," or shrank and disappeared, recalled one of the scientists, Thomas Doherty. "This was a surprising event."[22] The scientists then persuaded a surgeon to give nitrogen mustard to a man who was dying of a similar cancer. His tumors, too, shrank dramatically, though not permanently.

At the time, few doctors believed that cancer could be treated with drugs: Surgery and radiation were the only accepted treatments. Nonetheless, nitrogen mustard was soon joined by a host of other anticancer drugs. Because of these drugs, some kinds of cancer that were once invariably fatal, such as one type of leukemia (blood cell cancer) that strikes children, can now be cured some or even most of the time. Even when drugs do not completely stop a cancer, they can add years to cancer patients' lives.

More mysterious than microbe-caused diseases or even cancer were illnesses that struck the mind, causing people to behave strangely or hear and see things that did not exist. Until the 1950s, most people with severe mental illnesses such as schizophrenia and depression were institutionalized, often for life. Doctors and nurses in mental hospitals could do little for their patients except confine them, take care of their physical needs, and try to keep them from hurting themselves or others. This picture changed dramatically, however, when researchers discovered several drugs that had spectacular effects on severe mental illness.

## A Calming Drug

Finding any one of these drugs would have brought a scientist well-deserved fame, but an American psychiatrist named Nathan Kline helped to develop two of them. The first drug came from a plant called snakeroot, which had been used in India for thousands of years to treat a variety of conditions, including mental illness. In 1949 an Indian researcher, Rustom Vakil, reported that snakeroot lowered blood pressure. He also noted that it made patients relaxed and sleepy. Soon afterward, Western scientists isolated an extract of the plant and named it reserpine.

That was where Nathan Kline came in. In 1953 he was the director of research at Rockland State Hospital, a mental hospital in Orangeburg, New York. He needed an expensive piece of laboratory equipment and asked Squibb, an American drug company, for a grant to buy it. Squibb agreed to give him a grant if he would try reserpine on his patients to see whether it calmed them.

Kline tried reserpine on some seven hundred schizophrenic patients at Rockland. Excitedly he wrote to Squibb:

> Every ward that has used the drug has . . . reported a marked decrease in the decibel [noise] level, an increase in the cooperativeness of the patients,

### Joyful News

*In* Patenting the Sun, *Jane S. Smith describes the excitement with which Americans greeted the 1955 announcement that tests of the Salk polio vaccine had been successful and mass protection of children against the crippling disease could begin.*

"All over the country, people turned to their radios at ten o'clock to hear the full details. In Europe they tuned to the Voice of America broadcast. Judges suspended trials so that everyone in the courtroom could hear; department stores set up radio loudspeakers so shoppers could follow the news. Flushed by the first report that the vaccine worked, exuberant citizens rushed to ring church bells and fire sirens, shouted, clapped, sang, and made every kind of joyous noise they could. City councils and state legislatures postponed their regular business to draft resolutions congratulating Dr. Salk for his wonderful achievement."

*Jonas Salk produced the polio vaccine first distributed in the United States.*

## Prisons for Mental Patients

*Donald Robinson, in* The Miracle Finders, *quotes Frank J. Ayd's description of twentieth-century mental hospitals before drugs to treat severe mental illness were discovered in the 1950s.*

"Within the bare walls of isolated, overcrowded, prison-like asylums [mental hospitals] were housed many screaming, combative individuals whose animalistic behavior required restraint and seclusion. Catatonic patients stood day after day, rigid as statues. . . . Their comrades idled week after week, lying on hard benches or the floor, aware only of their delusions and hallucinations [imaginary experiences]. Others were incessantly restive [constantly restless], pacing back and forth like caged animals in a zoo. Periodically, the air was pierced by the shouts of a raving maniac. Suddenly, without notice, like an erupting volcano, a . . . schizophrenic bursts into frenetic [wild] behavior, lashing out at others or striking himself with his fists. . . . Nurses and attendants, ever in danger, spent their time protecting patients from harming themselves or others. . . . For lack of more effective remedies, they secluded dangerously frenetic [wildly excited] individuals behind thick doors in barred rooms stripped of all furniture and lacking toilet facilities. They restrained many others in [hand]cuffs and [strait]jackets or chained them to floors and walls. . . . These measures, barbaric and inhumane as they appear in retrospect, euphemistically called therapy, at best offered protection to patient and [hospital] personnel and a temporary respite from the most distressing symptoms of psychoses [severe mental illnesses]."

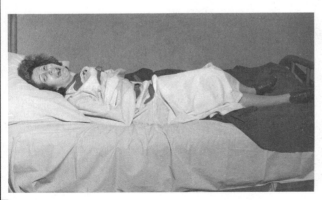

*Prior to the development of drugs to treat the mentally ill, patients such as this woman were restrained in straitjackets, handcuffed, or chained to walls and floors.*

*Nathan Kline administered reserpine to schizophrenic patients at his mental hospital and found the drug to have a calming effect.*

and decidedly less need for restraints, isolation, and seclusion.[23]

Even more important, the drug took the pain out of mental illness for the patients themselves. Kline described the effect of a reserpine injection on a schizophrenic woman who had been crying out in fear because she believed she was in hell.

> Visibly and measurably the mania [extreme excitement] began to ebb away. She was not suddenly cured of her delusion [false belief], but the terror had gone out of it. One had to question her closely now to draw out of her the fact that she still thought she was in Hades [hell]. And over the days that followed, the delusion itself began to fade as we continued the drug treatments.[24]

The most common treatment for mental illness at the time was psychotherapy, which consisted mostly of long talks with a psychiatrist. This kind of treatment worked well with mild depression or anxiety, but it was often ineffective with severe mental illnesses. Psychiatrists were startled to learn that something out of a bottle could succeed where all their talk had failed. Kline's claims were soon supported, however, by reports about another drug, chlorpromazine, that produced results just as dramatic as those of reserpine.

## Swinging the Pendulum Up

Kline next began looking for a drug to help people who were deeply unhappy or depressed. "If one compound could swing the emotional pendulum down," he reasoned, "then there should be another compound that could swing it up."[25]

Kline gained a vital clue to the compound he was seeking when he gave a lecture at the laboratories of another drug company, Warner-Chilcott, one day in 1956. After the talk a company scientist, Charles P. Scott, told Kline about some strange results he had gotten when he gave a drug called iproniazid to mice. Iproniazid was used to treat the lung disease tuberculosis. The drug, Scott said, made his mice unusually energetic and excited.

Kline thought iproniazid might be the "psychic energizer" he had been searching for. Looking back into the scientific literature, he found that some doctors who had given iproniazid to patients with tuberculosis had noted that it made the patients feel happier and more lively as well as helped them get well. Those doctors were not thinking about mental illness, though, so they had paid little attention to this observation. "The evidence had been right

*A patient sits in his mental health facility. The development of mental illness medication in the 1950s and 1960s improved the lives of many patients.*

under our noses," Kline said, "but everybody had missed it because nobody was looking."[26]

In November 1956, Kline began testing iproniazid on severely depressed patients. The results were just as spectacular as were the improvements seen with reserpine. Some patients who had hardly moved for fifteen years began sitting up, speaking, and eating almost normally.

Nathan Kline twice received the prestigious Lasker Award, in 1957 for his work with reserpine and again in 1964 for iproniazid. The second award citation stated,

Dr. Kline more than any other single psychiatrist has been responsible for one of the greatest revolutions ever to occur in the care and treatment of the mentally ill. Literally hundreds of thousands of people are leading productive, normal lives who—but for Dr. Kline's work—would be leading lives of fruitless despair and frustration.[27]

Reserpine, chlorpromazine, and iproniazid were just the first of the magic bullets against mental illness that were discovered in the 1950s. Some of the new drugs helped people with major mental illnesses. Others soothed, or tranquilized, normal people who were temporarily anxious or upset. In 1956, *Life* magazine called the new mental illness drugs "one of the most spectacular triumphs in the history of medicine."[28] By the 1970s, medication had replaced psychotherapy as the most common treatment in American psychiatry.

## New Challenges, New Hope

The search for magic bullet drugs continues today. New antibiotics are in fact needed, for many microbes have become resistant to the old ones. Resistance develops when, by chance, a microbe becomes able to make a chemical that destroys a drug or keeps it from harming the microbe. Microbes with genes for making this chemical survive and multiply while those that lack the protective genes die. Cancer cells become resistant to anticancer drugs in a similar way.

New drugs are also needed because new forms of disease sometimes appear. Perhaps the most challenging disease that doctors face today was first recognized in the early 1980s. It is AIDS (acquired immunodeficiency syndrome), which most researchers believe is caused by the human immunodeficiency virus (HIV). HIV destroys certain essential cells in the immune system, leaving people unable to defend themselves against microbes.

So far, there is no vaccine or cure for this fatal disease. Scientists around the world, however, are racing to discover both. They have developed vaccines that seem to protect monkeys or apes against viruses related to HIV, and some of these are being given preliminary tests in humans. Medical researchers have also created several drugs that greatly slow the advance of the disease.

The first drug to be used successfully against AIDS was azidothymidine, or AZT. This drug began to be used widely around 1987. Related in structure to thymine, one of the bases in RNA and DNA, AZT keeps the AIDS virus from reproducing by fooling it into using the drug rather than the real base when the virus makes copies of itself. Once a virus copy includes AZT, the copying process cannot be completed.

Scientists have learned that AZT works best when people start to take it soon after they have become infected with HIV. At that time their immune systems are not yet severely damaged, and they show no signs of AIDS-related illnesses. AZT also is most effective when taken in combination with other anti-AIDS drugs such as ddI (didanosine) or one of a new class of drugs called protease inhibitors. Protease inhibitors attack a chemical that HIV uses when it reproduces.

None of these drugs cures AIDS. The drugs can, however, give people with the disease years, even decades, of relatively healthy life. They may help many people with AIDS survive until a true cure can be developed. As Sandra Hernandez, director of the San Francisco Department of Health, says,

> For too long, people [with AIDS] have prepared themselves to die. We need to get people prepared to live, and live longer. . . . We've been grieving, and there has been some level of acceptance of death. But we shouldn't be accepting this anymore.[29]

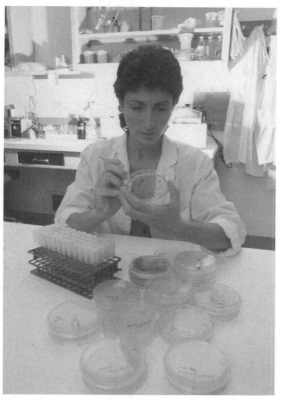

*A scientist in the process of engineering genes for the production of an AIDS vaccine. Currently drugs are available that can slow the progression of AIDS, but there is no cure for the disease.*

# Chapter 5
# Rebuilding the Body

## Windows into the Body

One day in 1895, a German woman stared in disbelief at a photograph of her hand. The hand's skin and muscles were reduced to a faint outline, but the bones stood out so clearly that she might have been looking at the hand of a skeleton. A band circled one of the skeleton's fingers, marking the wedding ring she wore.

The horrified housewife was the wife of physicist Wilhelm Röntgen. Röntgen had used his wife to test the effects of a form of radiation he had accidentally discovered. This radiation penetrated most materials, including the soft tissues of the human body, but was slowed by bone and stopped by metal. When a body, or part of a body, was placed between a source of the radiation and an unexposed photographic plate, the result was the sort of skeletal picture that startled Frau Röntgen. Because Röntgen did not know what this new radiation was, he called it X rays.

A week after Röntgen described his discovery in a scientific journal, his find was front-page news in Vienna, London, and New York. Newspapers called it a "marvellous triumph of science."[30] Doctors recognized immediately that the German physicist had given them a miracle: a way to see inside the body without cutting it open. The use of X rays to diagnose, or find the cause and nature of, medical problems quickly became widespread.

Amazing as they were, X rays had their limitations. They were excellent for show-

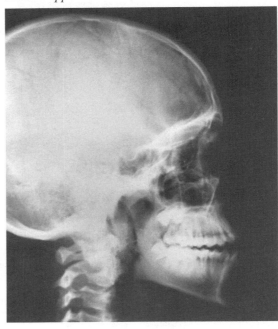

*When X rays, which penetrate soft tissue but are slowed by bone, pass through the body onto an unexposed photographic plate, a picture of the skeleton appears.*

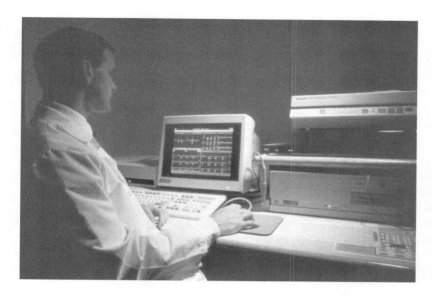

*The EEG detects the brain's electric currents through electrodes attached to the scalp. The electric impulses were originally scribed by a needle moving on a scroll of paper but are now recorded by computer (pictured).*

ing bones or metal objects, such as a bullet embedded in flesh or a coin that a baby had swallowed. They showed little of the soft tissues that made up vital organs such as the heart and brain, however.

In the first half of the twentieth century, doctors found a few ways to enhance the power of X rays. They could obtain good photographs of the digestive system, for instance, by making a patient swallow a substance containing the element barium, which blocked the rays. Many tissues, however, remained off-limits to X rays.

Doctors also sometimes looked inside the body with a device called an endoscope. Invented by Phillipe Bozzini in Germany in 1806, the endoscope was simply a hollow tube that could be inserted into a body opening. Doctors first used candlelight as their illumination when they peered down the tube. Later they used electric light, such as a flashlight.

Electricity provided other kinds of information about the hidden interior of the body. The electroencephalograph (EEG), developed in 1929 by a physicist who became a psychiatrist, used electrodes attached to the scalp to detect tiny electric currents produced by the brain. These electric impulses were translated into jagged black lines scribed by a needle moving on a scroll of paper. EEG patterns could reveal such conditions as epilepsy, in which abnormal discharges of certain nerve cells in the brain cause a person to have convulsions. A similar device, the electrocardiograph, shows the electrical pattern of the heartbeat and helps doctors diagnose heart disease. Both devices are still in use, though today a computer's monitor screen usually replaces the needle and paper.

## Modernizing Diagnosis

In the second half of the century, new diagnostic technology transformed the old workhorses of X rays and endoscopes into tools that seemed as miraculous as X rays had when they were first demonstrated. It also created new tools that provided unimagined views into the body.

First, in the 1950s, a new technology called fiber optics gave a whole new dimension to endoscopes. Optical fibers are flexible strands of glass, no thicker than a hair, that can conduct light almost the way a pipe carries water. One bundle of glass fibers in a modern endoscope carries light in, while another brings it back out. Fiber-optic endoscope tubes are much longer, narrower, and more flexible than the old tubes were. They can be inserted several feet into the body to give doctors a close-up look at the respiratory, digestive, or urinary tract.

Similarly, beginning in the early 1970s, computers revolutionized the old technology of X rays. British researcher Geoffrey Hounsfield found a way to make a computer combine X-ray pictures taken from many different angles into a single image that showed a "slice" or cross-section of the body. This new perspective allowed doctors to see organs that had been blocked by bones in conventional X-ray photographs. It could detect tumors and other abnormalities in the brain, for instance. Hounsfield's technique was called computerized axial tomography, usually abbreviated as CAT or CT. Most large hospitals have CAT scan machines today. Some can combine the slices to make a three-dimensional picture of part of the body.

## New Images

Some new diagnostic tools grew out of technology that did not exist before midcentury. One of these came into use because a Scottish scientist named Ian Donald asked a strange question: How is an unborn baby like a submarine? The answer was that both are surrounded by liquid, and both could be detected by echoes made when high-frequency sound waves struck them.

Technology called sonar had been developed during World War II to help ships locate enemy submarines. Sonar devices sent sound waves, too high in pitch (frequency) for the human ear to hear, through the water. If the waves struck a submarine or other solid object, they bounced back, making echoes. The echoes could be detected and translated into signals shown on a screen. Donald realized that a form of sonar could also be used to examine a fetus, or unborn baby, in a woman's fluid-filled uterus. The sonarlike images could reveal birth defects or other problems with the fetus's development.

Later this sonography, or ultrasound, technique was adapted to show tumors, fluid-filled growths called cysts, and certain other abnormalities inside the body. A variation of ultrasound is used to study blood flow. Computer processing combined with ultrasound can even be used to make moving pictures of the digestive system or other parts of the inner body. As Barbara Carroll of Stanford University Medical Center in California says, "The way things move or don't can often tell us when something's wrong."[31]

Another new diagnostic technology, which came into use in the 1980s, grew out of discoveries in atomic physics. Called magnetic resonance imaging or MRI, it is based on the fact that when exposed to a strong magnetic field and then hit by a burst of radio waves, the nuclei of certain atoms send out electromagnetic signals. A computer can translate these signals into images.

Unlike a CAT scan, MRI does not expose the body to X rays, which can be dangerous. MRI also produces clearer images than CAT. Doctors can use MRI to detect very small cancers or show blockages in blood vessels. "With MRI we are able to see brain structures we could never see before,"[32] says Harold L. Rekate, an expert on brain surgery in children.

## Cutting with Light

New diagnostic tools have been combined with other new technologies to greatly change surgery. Repeated CAT and MRI scans can now be made while surgery is going on, guiding the surgeon from moment to moment. Cutting tools and other surgical instruments can be inserted through endoscope tubes, allowing the surgeon to act on what he or she sees. New devices such as lasers can also be used with endoscopes.

Lasers are probably the most important of the new surgical tools. A laser beam is a beam of pure light, consisting of waves that all have the same wavelength and move in the same direction. It can be made hot enough to cut like a knife, sealing blood vessels as it goes and thus performing "bloodless surgery." Laser beams can be controlled and aimed so precisely that they can hit a target only one micron (about 0.00004 inch) across.

Lasers were invented in 1960 and began to be used in eye surgery soon afterward. Surgeons first used them to reattach the retina, the sensitive light-gathering tissue at the back of the eye, which sometimes becomes partly detached from the inside of the eyeball. Later, eye surgeons learned how to use lasers to remove cataracts, the cloudy lens (the part of the eye that focuses light) that sometimes develops with age. Some eye surgeons today use lasers to reshape the cornea, the clear covering on the front of the eye, thus correcting nearsightedness. Lasers are also

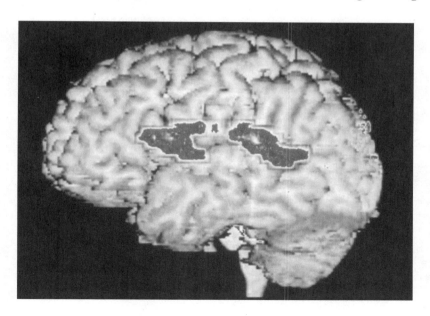

*The MRI was developed from the idea that the nuclei of some atoms emit electromagnetic signals when exposed to a magnetic field and bombarded by radio waves. The signals are translated into images, like this view of the brain.*

used in operations on joints, blood vessels, kidneys, and more. They provide an effective way of surgically removing several kinds of cancer.

## "Video Game" Operations

Either lasers or conventional cutting instruments can be combined with a fiber-optic endoscope called a laparoscope to create a versatile new tool for abdominal surgery. Beginning in the late 1980s, surgeons used laparoscopes to remove gall bladders (a small organ that helps in digestion), appendixes, and parts of the female reproductive system. These kinds of operations used to require making a large opening in the patient's abdomen, resulting in a long scar, a great deal of pain, and weeks of recovery time. With a laparoscope, however, the same operations require incisions (openings) so small that they can hardly be seen. Patients often leave the hospital a few hours after the operation and are back at work in a few days. This saves not only suffering but the high costs of a hospital stay and lost work time. "We are doing less invasive things" in surgery, says surgeon Arthur Gary. "It's benefiting the patients tremendously."[33]

In a laparoscopic procedure, gas is pumped into the abdomen so that the abdomen's outer wall is lifted away from the operating area. The laparoscope is inserted through a small incision, often in the navel. It produces images that the surgeon can see on a video screen. Lasers or other cutting tools are inserted through the endoscope or through other incisions. As the surgeon operates, his or her actions

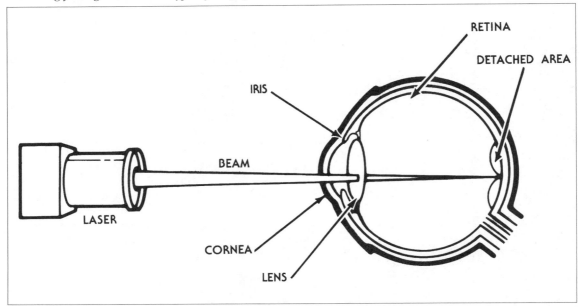

*A laser reattaches the retina to the eyeball. Because they are precise and make bloodless incisions, lasers were first used in eye operations, but they are increasingly being used in other types of surgery as well.*

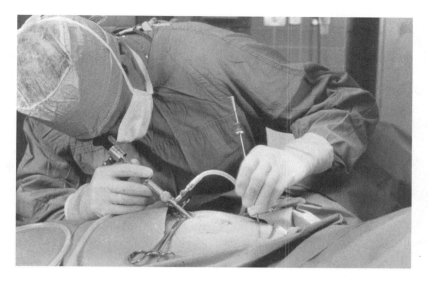

*A surgeon peers into a patient's abdomen during a laparoscopy. The laparoscope is inserted through a small incision and produces images on a video screen as the surgeon operates.*

are shown on the screen. Operating with the help of this technology is a little like playing a video game. In the future, computer technology called virtual reality may make the video images appear to be three dimensional. It may even give surgeons touch sensations like those they would have in a real operation.

## Robot Surgeons

A few surgeons have used robots to perform part of certain operations. In 1985, for instance, Yik San Kwoh of Long Beach Memorial Medical Center in California used a robot arm to help him remove tissue

### What Is an Image?

*In* A History of Medicine, *by Jenny Sutcliffe and Nancy Duin, radiologist (X-ray specialist) Bill Lees describes how computer technology has changed people's concept of what an image is.*

"The impact of computing has changed our concept of what constitutes an image. Once, an image was a physical entity, a pattern of shadows on a piece of film. Now, it is more likely to exist as a pattern of numbers held within a computer as a three- or many-dimensional model of the patient that can be manipulated and displayed in many different ways. Increasingly, imaging devices will be networked and their output electronically stored and distributed, potentially accessible anywhere in a country's health service."

samples from suspected tumors deep in the brain. "The robotic arm is safer, faster, and far less invasive than current surgical procedures,"[34] Kwoh said.

More recently, orthopedic surgeon William Bargar of Sacramento, California, used a different robot, which he calls Robodoc, to drill out a socket in the femur, or thighbone, for an artificial replacement hip joint. Robots like these, controlled directly by the surgeon or by a computer program prepared by the surgeon in advance, can make movements more steady and accurate than those of any human surgeon. Robodoc is said to be twenty times more accurate in drilling than a human doctor.

## From Sausage Casings to Saved Lives

At the same time that advances in diagnosis and surgery have allowed doctors to identify and remove diseased body parts, other advances have enabled them to replace some of those body parts with healthy ones. Some replacements are artificial; others are transplants of living organs.

Great strides have been made in creating artificial limbs. Providing artificial substitutes for internal organs has been more challenging, however. Today, artificial organs are used mainly to keep people alive until they can obtain transplants of living ones.

In 1938, just before World War II, a Dutch doctor named Willem Kolff made the first machine that successfully replaced a major human body organ. Kolff's device was intended to replace the kidneys, two organs near the lower back that filter wastes from the blood. When the

*Willem Kolff invented the first artificial organ—a machine that filtered impurities out of the blood when the kidneys failed.*

kidneys fail, poisonous wastes accumulate, causing death within a few days. Blood flowed from the patient through a tube into Kolff's machine, which did the filtering that the person's kidneys could no longer accomplish. Another tube sent the purified blood back into the patient. The first version of the machine used cellophane sausage casings for the filters.

Kolff first attached human beings to his machine in 1943. At first the device prolonged patients' lives by only a few days. Still, Kolff was sure he was on the right track. He told an interviewer later,

> When these patients were brought to me, they were mostly comatose, practically moribund [almost dead]. I saw them regain their consciousness. I saw them talk to their families. . . . Even when I lost them, . . . I knew that I had seen a temporary improvement. I was sure that in time I would get one who would be saved.[35]

During the time he was working on the artificial kidney, Kolff's life was in constant danger because he was resisting the Nazis, who occupied the Netherlands in 1940. He often hid members of the anti-Nazi underground in his hospital, for instance. Yet saving lives was more important to Kolff than politics. The first person whose life Kolff's artificial kidney saved, in 1945, was a woman who had helped the Nazis during the war. "People begged, 'Let her die,'" Kolff said later. "But no physician has the right to decide whether a patient is a good guy or not. He must treat every patient who has need of him."[36]

Improved versions of the artificial kidney, or dialysis machine, are still in use. Until recently, their users had to spend several hours in a hospital two or three times a week. Treatments were uncomfortable and left patients feeling weak; they were also very expensive. Now portable dialysis machines can be used at home, but they still serve mainly to keep people alive until donor kidneys are available for transplant.

## Artificial Hearts

Machines that take over the function of the heart have proved harder to develop than those that replace kidneys. The first such machine intended to keep humans alive on a day-to-day basis was invented by Robert Jarvik of the University of Utah. Called the Jarvik-7, the heart was made of plastic and a metal called titanium. It pumped blood by air pressure, using an outside source of compressed air.

In 1982 Barney Clarke, a sixty-one-year-old dentist, became the first person to be attached to a Jarvik-7. Clarke lived for 112 days. A later Jarvik heart recipient, William Schroeder, survived for twenty months. Life with a Jarvik heart, however, was far from normal. Patients were attached at all times to the bulky machines that provided the air to run the hearts. Far more serious, they often suffered damaging strokes when clots formed in their blood vessels and traveled to their brains, cutting off blood flow there and killing vital tissue. Improved models of artificial hearts now exist, but, like dialysis machines, they are used mainly for temporary support of people awaiting transplants.

*Dialysis machines like this one still function as replacements for the kidneys, but they are usually used temporarily until a transplant is available.*

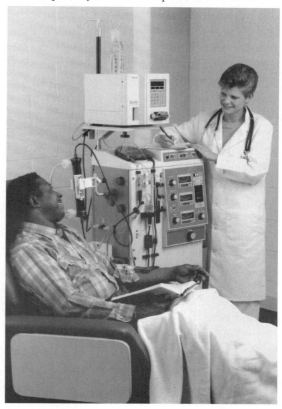

## "Transplanting" Blood

Legends claim that doctors in ancient Greece and China transplanted organs from one human being to another. No proof of successful human organ transplants occurred, however, until this century. The great bar to transplants was not the surgery involved but the fact that the body's defending army, the immune system, sees the lifesaving but alien new organ as an enemy invader and tries to destroy it. Transplants became common only after researchers found ways to blunt this attack.

Probably the first part of the human body to be transplanted safely was blood. Doctors occasionally had tried to transfer blood from one person to another before the twentieth century, and sometimes the transfusions had worked. More often than not, however, the people who received the blood died. Just after the turn of the century, an Austrian doctor named Karl Landsteiner found out why. Landsteiner discovered that when he mixed blood from two people, the cells in the blood often clumped, or stuck together. The blood then could no longer flow.

This reaction is caused by the immune system. Chemicals in the receiver's blood, now called antibodies, identify other substances called antigens on the donor's blood cells as foreign. The antibodies attach to the antigens. Cells in the immune system are drawn to these antigen-antibody combinations and destroy the donor cells.

Landsteiner discovered two antigens on red blood cells, which he called A and B. Some people's cells have only one of these antigens. Others have both or neither. People have antibodies to whichever antigen their own blood does not have. Thus, someone whose blood has the A antigen—a person with what Landsteiner called blood type A—can safely receive blood only from someone else who is also type A or who is type O (carrying neither A nor B antigen). People with antigen B, similarly, can receive only their own blood type (type B) or type O. Type AB people have both antigens and no antibodies to them, so they can receive blood from anyone. Type O people, on the other hand, have antibodies to both antigens, so they can receive blood only from other people with type O.

Later researchers learned that the system of antigens and antibodies in the blood is really much more complicated than this. Landsteiner's research, however, took the mystery out of transfusions' suc-

*Blood transfusions became possible early in this century, when it was discovered that different blood types existed and some were compatible.*

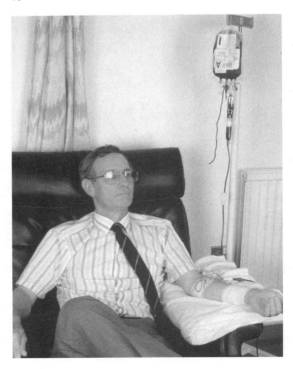

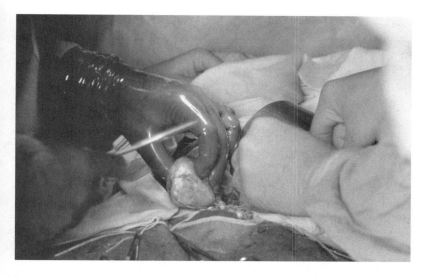

*Kidney transplants, shown here, were the first successful organ transplants. Since a kidney could be donated by a relative with similar genes, the recipient's immune system was less likely to reject the kidney.*

cess and failure. Using what he had learned, hospital workers learned to mix blood samples from prospective donors with those of recipients to see whether the cells clumped. If this did not happen, the transfusion would probably be a success. Landsteiner's discovery, as well as others that made it possible to preserve blood and blood products, made transfusions possible on a wide scale during World War II and saved many lives.

## New Organs for Old

Kidneys were the first body organs to be transplanted successfully. This was partly because, unlike other organs such as hearts, they could come from living donors. (Most transplants come from young, healthy people who are killed suddenly in such things as car accidents.) Everyone has two kidneys, but a person can lead a healthy life with only one. Thus, if someone needs a new kidney, a member of his or her family can be asked to donate one. The more closely related an organ donor and recipient are, the more genes they will have in common and the less likely it is that the recipient's immune system will attack, or reject, the new organ. From this standpoint, transplants between identical twins are ideal, since all their genetic information is the same. Indeed, the first successful human kidney transplant, in 1954, involved identical twin brothers.

In the 1960s, researchers discovered drugs that suppressed the immune system and therefore kept transplanted organs alive in their new location. The drugs were not always effective, and they had many undesirable side effects, including leaving patients open to attacks by microbes and cancer cells. They worked well enough, however, to allow surgeons to transplant kidneys and a few other organs from dead donors to unrelated recipients with fair success. Some surgeons then felt ready to take on the most dramatic task of all—transplanting a human heart.

The first person to receive a heart transplant was a fifty-four-year-old South African grocer named Louis Washkansky.

> ### Heart Transplant Hysteria
>
> *In* To Mend the Heart, *by Lael Wertenbaker, heart transplant surgeon Norman Shumway recommends a cautious attitude toward this and other "medical breakthroughs."*
>
> "The initial enthusiasm, approaching hysteria, which greeted the first clinical cardiac [heart] transplants now seems to have been replaced by a generally pessimistic outlook. Both reactions are probably inappropriate.
>
> The journalists' 'medical breakthrough' rarely results in a radical change in patient care. It is, rather, the cautious clinical application of information gained in the laboratory that results in the gradual development of new therapeutic [treatment] methods."

His new heart came from Denise Darvall, a twenty-five-year-old woman who was hit by a car when she and her mother crossed a street. Surgeon Christiaan Barnard transferred Darvall's heart to Washkansky in a hospital in Capetown, South Africa, on December 3, 1967. The procedure made headlines all over the world; *Newsweek* magazine called it the "Miracle in Capetown." Washkansky lived for just eighteen days, however, before dying of a lung infection.

Barnard's next heart transplant patient, Philip Blaiberg, lived a year and a half, but during most of that time his health was poor. Other surgeons who tried heart transplant operations had similar results: Sooner or later, usually sooner, the recipient's immune system destroyed the donor heart. After a few years, heart transplants began to fade from popularity as doctors and patients realized that they did little to prolong life or health.

In the 1980s, however, researchers discovered a new drug called cyclosporine, which comes from a fungus. Cyclosporine suppresses part of the immune system, protecting a transplanted organ without leaving the body completely defenseless. Thanks to cyclosporine and other drugs, transplants of kidneys, livers, and even hearts have now become almost routine. Some people have lived with transplanted hearts for twenty years.

Looking toward the future, some doctors think the day will come when, instead of receiving transplanted organs, people whose organs have failed will be able to grow new ones. These "organoids" will be made from healthy cells, either taken from the recipients' own bodies and stored or taken from compatible donors. A combination of growth-promoting substances and, perhaps, genetic engineering will shape the cells into new organs. "If salamanders can regenerate lost parts," says researcher John Thompson, "it's not too farfetched to think that humans can."[37] If this day comes, doctors may reach the goal of being able to replace any part of the body that is needed.

# Chapter 6
# The Information Revolution

They calculate phone bills and the shape of airplane wings. They weave together encyclopedias in pictures and sound. They count stars and subatomic particles. They bring dinosaurs to life in movies and show doctors three-dimensional pictures of the living human body. They do humble jobs hidden within car engines and kitchen appliances. They are computers, and they are everywhere.

Modern science depends on computers. Analyzing the inside of an atom or a living cell is too complicated to do without them. Only computers can move a telescope with the accuracy today's astronomers need or collect and interpret information from satellites in space.

Computers have also given scientists a new way to experiment. By putting together data from observations and rules from theory, scientists can create a simulation program that shows how the parts of an atom, cells in a colony of bacteria, or storm systems behave. Simulations can help scientists try out new ideas and plan future experiments.

Computers have opened up new areas of research as well. One of these is artificial intelligence, or ways to make computers think as people do.

The basis of modern computers is electronics—the ability to precisely control the flow of electromagnetic energy. In the late nineteenth century, when Crookes discovered cathode rays and Thomson identified electrons as the fundamental particles of electric current, these findings seemed to have little practical use. Both, however, were to play important parts in the twentieth-century electronics revolution.

## Radio Waves

At the same time these discoveries were taking place, inventors such as Guglielmo Marconi were starting to use electromagnetic waves to carry messages. Their "wireless telegraph"—later called radio—depended on the ability to detect electromagnetic signals from hundreds or even thousands of miles away.

The first devices used to detect radio waves were part electrical and part mechanical. One device, called a coherer, consisted of a tube filled with small bits of metal. When a radio wave entered the tube, the bits clung together, allowing current to flow through the conducting metal and click a sounder. Unfortunately, the tube then had to be hit by a hammer to loosen the bits again in preparation for the next radio signal. This meant that radio messages had to be sent slowly. To speed

*Guglielmo Marconi was one of the first inventors to use electromagnetic signals, or radio waves, to carry messages.*

up transmission, radio engineers needed a detector that would respond instantly to radio waves and not have to be reset.

In 1905 a British physicist named Ambrose Fleming took a major step toward inventing such a detector by making a glass tube much like a light bulb. The tube had two electrodes through which it could be connected to an electrical circuit. One electrode, called the cathode, was much like the filament of a light bulb. When the circuit was turned on, the cathode sent a stream of electrons across the airless interior of the tube. The electrons were attracted to the other electrode, or anode, which was a metal plate that had an opposite, positive charge.

Radio signals consist of waves whose motion produces an electric current that alternates, or moves back and forth, many times a second. As it moves, the current changes between being positively and negatively charged. The problem for radio engineers was to turn the alternating current into a current that still moved back and forth with the motion of the waves but always had the same charge and so could be kept flowing toward the same electrode. This is called a direct current. Unlike alternating current, direct current can be easily changed to sound waves that can be heard in an earphone when a signal is received.

Fleming discovered that if the alternating current from a radio signal is applied to the anode in a Crookes tube, the negative part of the current will not flow because it is the same as the negative charge on the cathode, and like charges repel or block each other. Only while the alternating current is positive does it flow across the tube. The tube therefore turns the alternating current into a direct current that still contains the pattern of the radio signal.

In 1907 another inventor, Lee de Forest, added to Fleming's tube a bent wire or "grid" that carried a separate current, such as that produced by a radio signal. The grid current could be used to control the flow of the main tube current, in effect increasing, or amplifying, the signal by transferring its pattern of variation to the larger current.

By World War II the work of other electronics inventors such as Edwin Armstrong had built on these earlier inventions to create a new radio broadcast industry. The principles of electronics had been turned into a practical technology. During the war, the need to develop more sophisticated radio and radar systems gave a further boost to electronics.

## Global Communications

Electronics is one root of today's information age. The other is global communication. In 1945, as World War II was ending, communications researcher and science fiction writer Arthur C. Clarke made a startling proposal. Suppose, he said, the recently developed science of rocketry were used to put a satellite in a special orbit about 22,300 miles above the earth's equator? At this height, the speed at which the satellite was moving would exactly match the speed at which the earth rotates. As a result, the satellite would always stay above the same spot on the earth. The satellite could relay, or rebroadcast, any radio transmission beamed up to it, sending the transmission either to another satellite or to another spot on the earth. In the 1960s Clarke's vision came true. Today, satellites relay radio, telephone, and television signals around the world almost instantly.

Once satellites made worldwide communication possible, a torrent of information began to flow around the world. Ordinary electrical switches could not handle this growing traffic. Members of the science and business communities needed a way to sort, analyze, and process all the data being generated and transmitted. Fortunately, the computer offered a brain for the world's new electronic nervous system.

*A ground test for the world's first passive communications satellite in 1960. Satellites provide worldwide communication by relaying radio transmissions to other satellites or to locations on the earth.*

## Number Crunchers

World War II, which produced so many other advances in technology, also triggered the birth of the modern electronic computer. Governments involved in the war had an urgent need to break enemy codes, design new airplanes, and calculate the flight path of artillery shells. Early code-breaking machines and similar programmable calculators suffered from two problems, however. First, they used thousands of mechanical electric switches, which slowed their action. Second, they were not very flexible. They could be programmed only to run through a series of calculations in a given order. They could not deal with the need to do different things under different circumstances.

In 1946, a new type of calculating machine called the ENIAC (Electronic Numerical Integrator and Calculator) was unveiled. It used vacuum tubes to provide electronic switching, which made it thousands of times faster than mechanical calculators—at least until one of its nearly eighteen thousand tubes blew out. Like its wartime predecessors, it was huge, filling most of a large room.

ENIAC had one major shortcoming, as its designers, John Presper Eckert and John Mauchly, noted:

> No attempt has been made to make provision for setting up a problem automatically. This is for the sake of simplicity and because it is anticipated that the ENIAC will be used primarily for problems of a type in which one setup

---

### The First Computer Bug

*Charlene Billings, in her biography* Grace Hopper: Navy Admiral and Computer Pioneer, *repeats Hopper's story of the origin of the term* bug, *meaning a computer program error.*

"'In the summer of 1945 we were building the Mark II [computer]; we had to build it in an awful rush—it was wartime. . . .'

All the windows were open because it was a hot summer and there was no air-conditioning to relieve the oppressive heat.

Suddenly the Mark II stopped. After a while, the crew found the relay [switch] that had failed. Inside the relay was a moth that had been beaten to death by the relay.

Grace says, 'We got a pair of tweezers. Very carefully we took the moth out of the relay, put it in the logbook, and put scotch tape over it. . . . From then on if we weren't making any numbers, we told [the boss] that we were debugging the computer.'"

ENIAC, *an electronic calculator, required workers to set hundreds of switches and plugs to solve a problem.*

will be used many times before another problem is placed on the machine.[38]

To solve a problem using ENIAC, a team of workers had to set hundreds of switches and plugs that corresponded to the program instructions and data numbers. When they wanted to tackle a different problem, they had to set the switches all over again.

The EDVAC computer, put in use in 1947, improved this situation. It stored program instructions in its memory along with the data needed to solve a problem. Instead of setting switches for each problem, workers could feed the computer new program instructions encoded on tape or punched cards.

In the late 1940s, most people thought computers were useful only for large military and government operations. Mauchly and Eckert, however, began to change this perception when they produced UNIVAC, the Universal Automatic Computer. UNIVAC treated watchers of CBS-TV news to a successful prediction of presidential election results in 1952, for instance. UNIVACs and competing computers from IBM (International Business Machines) soon started to appear in large businesses. These computers were easier to use than their predecessors, but they still had to be programmed by highly trained specialists.

## Making Computers Easier to Use

If someone wanted an early computer to add two numbers, the computer had to be told to fetch each number from its storage

*Computers like UNIVAC (top) needed complicated instructions to carry out operations. Grace Hopper (bottom) alleviated the problem by inventing programs that translated instructions into commands the computer could execute.*

location, add the two together, and then place the result in a third location. More complex operations, such as multiplication or square roots, required dozens of instructions. Data had to be moved back and forth between the main processing unit and storage locations. Very few people knew how to create these detailed instructions.

Starting in 1951, mathematician Grace Hopper made life easier for computer programmers by inventing programs called compilers. Compilers let computers translate instructions written in human symbols (such as letters and numbers) into machine instructions the computer can execute. Hopper also headed a committee that created a "language" for programmers called COBOL (Common Business-Oriented Language). COBOL was almost like English, yet—with a compiler's help—a computer could also "understand" it. Many large business computers still use COBOL.

**82** ■ TWENTIETH-CENTURY SCIENCE

Several computer languages have been invented since, including some designed especially for scientists.

As designers developed computers capable of solving more kinds of problems, their programs grew longer, more complicated, and more likely to contain errors. Program designers looked for ways to organize and simplify computer programs. One technique they used was breaking programs into smaller units called subroutines. Each subroutine took care of one data-processing task. If a business used a program with subroutines and the office bought a new computer printer, for example, only the commands in the program's printing subroutine would have to be changed. Most programs today have hundreds of subroutines (sometimes called procedures or functions). These subroutines can be moved fairly easily from one program to another.

More recently, designers have carried program organization a step further by using what are called objects. An object is a "package" that contains all the subroutines or functions related to a particular task and all the data the computer needs to carry out that task. An object called VCR, for example, might contain functions for starting an animated graphic, stopping it, pausing, going forward, and rewinding. Programmers can add features to objects more easily than to subroutines.

## The Incredible Shrinking Computer

While computers were getting easier to program, they were also getting smaller and more powerful. Computers started

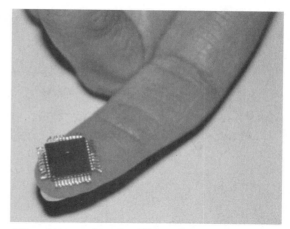

*Microchips condensed millions of circuits into a tiny space and consequently shrank the size of a whole computer.*

shrinking in the early 1970s, when builders began to use integrated circuits, in which many electronic parts are put together in a small space, instead of transistors. (Transistors, in turn, had replaced vacuum tubes in the late 1950s.) The resulting "minicomputers" were only the size of a refrigerator instead of being as large as a whole room, like their ancestors.

A few years later, inventors created the microprocessor, a complete computer processing unit on a chip. A chip is a wafer of silicon and other materials, about an inch square. It can hold thousands or even millions of circuits. Microprocessors made it possible to shrink computers a great deal more—but at first, few people could see any reason to do so.

In 1975, officials at computer giant Hewlett-Packard (HP) told a young college dropout, Stephen Wozniak, that they were not interested in his plan to build microcomputers (small computers that used microprocessors). Wozniak and his friend Steve Jobs built the machines anyway, starting what became Apple Computer

Company. No bank would finance a business that had no track record and produced such an untested product, so to raise enough money to go into production, Jobs sold his Volkswagen bus and Wozniak parted with his fancy HP programmable calculator. In 1977, the company's first year of operation, Apple sold $775,000 worth of computers. Four years later, its total sales had risen to $335 million.

Suddenly, for the first time, computers were something that ordinary people could afford, find space for, and—just possibly—learn to use. Seeing how successful the typewriter-size Apples were, other computer companies, including giant IBM, followed Apple Computer's lead. Soon budding entrepreneurs and knowledgeable amateurs were putting together their own computer systems in garage or basement workshops. Microcomputers began appearing everywhere.

The interest in microcomputers spawned an industry that went far beyond construction of the computers themselves. For example, microcomputer enthusiasts needed a program to run BASIC, a simple computer language that could be easily adapted to small computers. An enterprising young man named Bill Gates started a small company called Microsoft to fill this need. Gates is now the richest person in the United States.

As the 1980s progressed, some microcomputers shrank still further in size. Programs designed for these machines also became easier to use. Today, a preschool child can run some of them.

## A Multimedia World

In 1977, the head of a large computer manufacturing company declared, "There is no reason for any individual to have a computer in their home."[39] Today, on the contrary, computers are almost as com-

*Computers gradually became smaller, less complicated, and more widely available; now, even a young child can operate one easily.*

*While at a café, a patron explores the World Wide Web, a system that enables computers to access hypertext, multimedia, communications, and the Internet.*

mon in homes as television sets. People use them for business, home finances, schoolwork, games, and more.

Early computer programs were good at working with numbers but could do little with text or pictures. However, computer researchers Ted Nelson and, later, Douglas Englebart worked out a way for computers to help people connect ideas and images. It was called hypertext.

In a hypertext system, each document (such as a report, an article, or an encyclopedia entry) contains key words and phrases that are highlighted on the computer screen. When a user selects one of these highlighted areas, he or she is shown a related document or picture. For example, a student learning about Africa can choose different spots on a map of that continent to see descriptions and pictures related to each place. Today, a related technology called multimedia adds video and sound to the hypertext mix.

Meanwhile, users of university computers were developing a network by which they could send messages (electronic mail) to one another using telephone lines. Electronic bulletin boards or newsgroups also became popular. Users of bulletin boards could type in messages on a particular topic and read and respond to messages left by others. The network connecting universities and research labs became known as the Internet. Today, people from all walks of life and all parts of the world use this network to communicate with one another.

In 1990, researchers at CERN, a giant physics research institute in Switzerland, developed a way for users to make hypertext files available on the Internet. Their system is now called the World Wide Web (WWW). Like the Internet, it is open to any computer user with the right communication equipment and software. The World Wide Web brings together hypertext, multimedia, and communications and provides an easy-to-use gateway to the Internet. An ever-growing variety of individuals, businesses, and institutions have

> **Putting Computers in Perspective**
>
> *The World Wide Web and the Internet are exciting tools. However, as computer critic Clifford Stoll, in his book* Silicon Snake Oil, *points out:*
>
> "There's a relationship between data, information, knowledge, understanding, and wisdom.
>
> Our networks are awash in data. A little of it's information. A smidgen of this shows up as knowledge. Combined with ideas, some of that's actually useful. Mix in experience, context, compassion, discipline, humor, tolerance, and humility, and perhaps knowledge becomes wisdom.
>
> Minds think with ideas, not information. No amount of data, bandwidth [amount of data being moved], or processing power can substitute for inspired thought."

set up "home pages" on the Web. These computer sites feature everything from scientific papers to announcements of new products, ongoing fantasy games, and pictures of people's families or pets.

## Machines That "Think"?

In 1956, when computer science was in its infancy, a group of researchers organized a remarkable conference at Dartmouth University. The conference was supposed to

> proceed on the basis of the conjecture [theory] that every aspect of learning or other feature of intelligence can in principle be so precisely described that a machine can be made to simulate it.[40]

This goal was what one of the conference leaders, John McCarthy, called artificial intelligence, or AI.

At first, AI researchers were optimistic that they could simulate human thought. Early AI programs played chess or performed other specialized tasks quite successfully. These programs did not have the kind of general-purpose intelligence that people possess, however. A program might be better than most human chess players, but it did not have the slightest idea how to read a book or cook a meal.

The problem is that people can draw on a vast storehouse of facts about the way the world works. At present, computers have no such knowledge base. As John McCarthy says:

> Imagine a traveler who wants to fly from Glasgow [Scotland] to Moscow via London. Now imagine he has some facts about needing a ticket and the effect of flying from one place to another.
>
> Many programs can be made to reason that if he flew from Glasgow to London and London to Moscow, he'd be in Moscow. But what happens if he loses his ticket in London? You'll no

*A shopper looks for a home computer. Computer scientists are researching ways to expand computers' reasoning skills, making them more adaptive and analytical like those of humans.*

longer be able to say that the original plan will work. But a plan that involves his buying another ticket should work. Not one of the existing applied programs allowed for this kind of elaboration of a situation's conditions.[41]

One long-term AI project is attempting to give a computer program a vast storehouse of data about common situations from daily life, such as the fact that if you are at a restaurant and your food arrives at the table cold, you are likely to leave a smaller tip for the waiter. It is unclear whether such a "database of common sense" can be used to give computer programs the flexibility that human beings demonstrate every day.

Fortunately, many practical applications of AI do not require such a massive undertaking. AI programs called expert systems have limited knowledge bases that focus on a specific subject. Such a program is given a set of rules about a subject, derived from the way human experts think about that subject. As a simple example, a computer might be given the following rules to help it identify a sea creature:

> IF it is huge THEN it is a whale
> IF it is NOT huge THEN it is not-a-whale
> IF it is not-a-whale AND (it blows a jet of water OR it has a blowhole OR it has a horizontal tail) THEN it is a mammal
> IF it is not-a-whale AND (it has no blowhole OR it has a vertical tail) THEN it is a fish
> IF it is a mammal AND it has a pointed snout THEN it is a dolphin[42]

and so on. . . .

When an expert system is given a situation to analyze, it compares the elements of the situation with its "rules base" and finds the rule or rules that apply. It then uses those rules to reach conclusions about the situation. Expert systems have been used successfully to identify diseases, schedule airline flights, decide whether to approve bank loans, and design computer networks.

## Neural Networks

Other researchers have attempted to produce artificial intelligence by creating systems that start with little or no knowledge but can learn, somewhat the way the human brain does. In 1982, for example, John Hopfield of the California Institute of Technology wrote a computer program that he called a neural network. A neural network has many separate parts that act something like individual brain cells, or neurons. Each computer "neuron" can receive input from nearby neurons or a "stimulus" from outside the system.

To begin with, a stimulus is sent to the network, which calculates a response. The programmer decides what the correct response should be. If the response the network made is correct, nothing changes. If it is not, the network changes some of the settings of its neurons and tries again. Over time, the network becomes "trained" to give the correct response. What is remarkable is that, unlike the case with the expert system, the programmer does not tell the computer how to do something. The computer figures it out for itself.

AI researchers have even borrowed ideas from biology. For example, they have created computer programs that eat, grow, mate, and reproduce like living things. These programs carry encoded information that they use to attempt to solve problems, such as sorting data or finding paths on a map. The programs that do a better job are allowed to reproduce and pass their electronic "genes" on to their "offspring." This genetic approach to AI is starting to show promise.

Computers will surely become even smarter, easier to use, and more widespread in the future. People's relationship with them will become closer, providing both rewards and challenges undreamed of today.

# Chapter 7

# An Unpredictable Future

When times change rapidly, people try to prepare themselves for the future by predicting what they think will happen. Predicting the effects of new discoveries in science and technology has always been hard, however. The difficulty shows clearly in the way some predictions about the twentieth century have turned out.

## The Perils of Prediction

Some predictions about twentieth-century science and technology were accurate. In the late nineteenth century, for instance, French author Jules Verne wrote novels about submarines, airships, and a trip to the moon. None of these things existed when Verne described them. Imagining submarines and airships was not too hard, though, because inventors were already working with experimental models.

Other predictions went wildly wrong. Some were too ambitious. Almost since the airplane was invented, futurists (people who try to predict the future in a scientific way) have predicted that people would soon commute between home and work in small personal aircraft. That has not yet happened. Neither has the 1962 prediction of rocketry pioneer Wernher von Braun, who was sure that by 1984 human beings would have orbited or, more likely, landed on Mars.

On the other hand, some predictions of the future were not ambitious enough. In 1938, for example, an article in *Fortune* magazine stated:

> At present few scientists foresee any serious or practical use for atomic energy. They regard the atom-splitting experiments as useful steps in the attempt to describe the atom more accurately, not as the key to the unlocking of any new power.[43]

Similarly, Lee de Forest, a pioneer in the development of the radio, said in 1926 that although television might work from a technical point of view, "commercially and financially I consider it an impossibility."[44]

Entire technologies exist now that hardly anyone saw coming. For example, some science fiction stories of the 1950s featured computers as big as cities. These giant machines made decisions for everyone and, in effect, ruled the world. No one, however, predicted that computers would become small enough to hold in a person's hand, easy enough for a child to use, and cheap enough for most people to afford.

One reason accurate prediction is so difficult is that it is hard to anticipate the

*Alexander Graham Bell places the first long-distance call from New York to Chicago in 1892. People at that time could not anticipate the impact inventions like the telephone would have on the future.*

possible uses for a new idea or invention. According to one story, when Alexander Graham Bell's new telephone was demonstrated to a city mayor, the mayor commented approvingly that he was sure that someday every city would have *one*. People had to broaden their point of view before they could realize that individuals could use the phone to keep in touch with one another. Similarly, radio was seen at first as a "wireless" improvement on the telegraph, not as a medium that could broadcast messages to millions of people at a time.

It is also hard to guess all the ways an invention will affect people's daily lives. For example, the New York World's Fair of 1939–1940 had as its theme "The World of Tomorrow." Visitors to the fair could ride on suspended cars through Futurama, a "city of 1960" complete with soaring skyscrapers and spacious highways. When 1960 actually arrived, broad interstate highways indeed crisscrossed the land. Instead of moving people into high-rise cities, however, the automobile had spread them around the countryside.

## Reductionism

Failure to predict the future accurately is not due simply to human limitations, however. Scientists have discovered that unpredictability is a fundamental feature of nature itself.

In the seventeenth century, Isaac Newton showed that the motion of every body in the universe followed a few simple laws. About a hundred years later, the French astronomer-philosopher Pierre de Laplace suggested that a single sufficiently advanced intelligence

> would embrace in the same formula the movements of the greatest bodies of the universe and those of the lightest atom; for it, nothing would be uncertain and the future, as the past, would be present to its eyes.[45]

Scientists ever since, entranced by the beauty and explanatory power of natural law, have tried to find a few basic laws that would explain such complex subjects as weather, economics, and the human body. For example, Donald Frederickson, then director of the National Institutes of Health, said in 1982:

> The reduction of life in all its complicated forms to certain fundamentals that can then be resynthesized [recombined] for a better understanding of man and his ills is the basic concern of biomedical research.[46]

This belief that any natural system can be summarized in or reduced to a few general laws is called reductionism. Reductionism has dominated twentieth-century science. Today, however, more and more researchers are discovering that nature cannot be explained in such a simple way.

## Atomic Uncertainty

The first challenge to reductionism came from physics. In 1927, German physicist Werner Heisenberg pointed out that there would always be a basic uncertainty in measuring the tiny particles that make up atoms.

Science writer Heinz Pagels illustrates this idea with a comparison:

> If you look at a tomato seed on your plate, you may think that you have established both its position and the fact that it is at rest. But if you try to measure the location of the seed by pressing your finger or a spoon on it the seed will slip away. As soon as you measure its position it begins to move.[47]

To Heisenberg this discovery also meant that scientists are inescapably part of what they observe. He wrote:

*Isaac Newton demonstrated that the motion of every body in the universe followed a few basic laws. His idea contributed to the trend in twentieth-century science toward reductionism.*

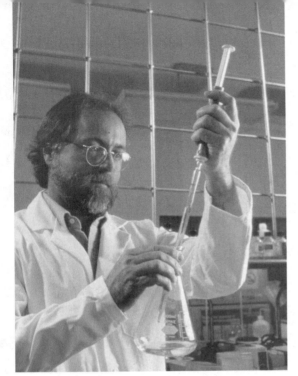

Natural science does not simply describe and explain nature; it is part of the interplay between nature and ourselves; it describes nature as exposed to our method of questioning.[48]

Put another way, the kind of answers that scientists get from nature depends partly on the way they ask their questions.

## A Chaotic World

Many scientists thought the uncertainty Heisenberg pointed out was unimportant. After all, it applied only to the tiny particles inside an atom. They assumed that larger systems, such as rainstorms or herds of deer, would still be governed by the kind of simple laws that reductionists sought. Computers seemed to be the perfect tool for finding such natural laws. Indeed, some researchers came to believe

*Part of the uncertainty of science is due to the fact that scientists cannot simply observe nature objectively; the observation itself affects their findings.*

### Future Trends

*In their 1982 book* Megatrends, *futurists John Naisbitt and Patricia Aburdene suggest ten ways in which they think society and the economy will change in the future.*

1. Industrial Society → to Information Society
2. Forced Technology → to High Tech/High Touch
3. National Economy → to World Economy
4. Short Term → to Long Term
5. Centralization → to Decentralization
6. Institutional Help → to Self-Help
7. Representative Democracy → to Participatory Democracy
8. Hierarchies → to Networking
9. North → to South
10. Either/Or → to Multiple Option

that anything in nature could be predicted by a sufficiently elaborate computer model. It was a computer model, however, that revealed why such models can never work perfectly.

In 1960 a meteorologist named Edward Lorenz made the first successful computer model of the weather. The model was simplistic, but its output showed air masses and winds interacting in a realistic way. Perhaps if his formulas could be made detailed enough, Lorenz thought, reliable weather forecasting would become possible.

One day in 1961, Lorenz decided to have his program predict weather patterns over a longer period of time than he had yet tried. To save time, he typed in numbers that his computer had already generated halfway through its previous run. Since he was using the same formulas as before, he thought the computer would simply retrace its steps and produce the rest of its previous output before going on to new territory. To his surprise, however, he found that the output almost immediately began to diverge, or move away from, its previous values. It soon ceased to resemble the output of the previous run in any way.

It turned out that the numbers Lorenz had typed were accurate to only three decimal places because the computer had rounded them before printing them out. The numbers actually stored inside the computer from the previous run, however,

*Even modern meteorological equipment like this cannot account for the complexity of natural phenomena.*

had six decimal places. That tiny difference produced a wildly different output. This was not in keeping with the expectations of reductionism, which assumed that output would change bit by bit according to formula as the input numbers were varied.

Weather proved to be one of a class of "chaotic phenomena" that are very sensitive to tiny changes in starting conditions. Even the paths of planets have proved to be subject to the variations of chaos over a long enough period of time. In short, nature is much more complex and less predictable than Newton and his followers thought.

Like Heisenberg's discovery of uncertainty in the measurement of subatomic particles, Lorenz's chaos theory does not mean that scientists cannot learn how nature works. It does mean, however, that the details of nature and the way those details interact are just as important as the general laws that describe systems. The discovery of chaos shows that people must be very cautious about altering a system because small changes can have large and unexpected effects. This is true whether the system is part of nature or of human society—which is itself a part of nature.

Twentieth-century science and technology made major alterations in both natural and social systems, often without considering the consequences, and the results were always surprising and sometimes disastrous. Paul Gray, then president of Massachusetts Institute of Technology, pointed out in 1992 that "a paradox of our time is the mixed blessing of almost every technological development."[49] The twenty-

---

### The Rewards of Science

*In* The Demon-Haunted World, *astronomer and popular science writer Carl Sagan explains how science will reward young people who keep working with it.*

"Because science carries us toward an understanding of how the world is, rather than how we would wish it to be, its findings may not in all cases be immediately comprehensible or satisfying. It may take a little work to restructure our mindsets. Some of science is very simple. When it gets complicated, that's usually because the world is complicated—or because we're complicated. . . .

But when we pass beyond the barrier, when the findings and methods of science get through to us, when we understand and put this knowledge to use, many feel deep satisfaction. This is true for everyone, but particularly for children—born with a zest for knowledge, aware that they must live in a future molded by science, but so often convinced in their adolescence that science is not for them."

*The vast developments in science in the twentieth century, such as the computers shown here, have changed society dramatically. Science will no doubt usher in even greater changes in the next century.*

first century is sure to offer other mixed blessings with even greater and more unpredictable consequences.

Should people then stop trying to predict the future, or should scientists and inventors stop trying to shape it with their discoveries? Not at all. Change is a fact of nature, and inventiveness and creativity are part of the human spirit. People need to find ways not to avoid change, but to plan for it, live with it, learn from it, and even enjoy it. Doing so is essential, because the next century is sure to bring change that no one today can dream of, any more than people in 1900 could imagine all the changes that the twentieth century has brought. As American science fiction writer Robert Heinlein has said about predicting the future,

> A "nine days' wonder" is taken as a matter of course on the tenth day.
>
> A "common-sense" prediction is sure to err on the side of timidity.
>
> The more extravagant a prediction sounds, the more likely it is to come true.[50]

# Notes

## Introduction: Science Shapes a Century

1. Quoted in Douglas McKie, *Antoine Lavoisier: Scientist, Economist, Social Reformer.* New York: Henry Schuman, 1952, p. 3.
2. Trevor I. Williams, *Science: A History of Discovery in the Twentieth Century.* Oxford: Oxford University Press, 1990, p. 8.
3. Quoted in Frederick Lewis Allen, *The Big Change.* New York: Harper and Brothers, 1952, p. 3.
4. Quoted in *The Colonial Overlords: TimeFrame AD 1850–1900.* Alexandria, VA.: Time-Life, 1990, p. 163.
5. Williams, *Science*, p. 17.

## Chapter 1: Inside the Atom

6. Quoted in Peter Vansittart, ed., *Voices 1870–1914.* New York: Franklin Watts, 1985, p. 194.
7. Quoted in Sean M. Grady, *Marie Curie.* San Diego: Lucent Books, 1992, p. 34.
8. Quoted in Lennard Bickel, *The Deadly Element: The Story of Uranium.* New York: Stein and Day, 1979, p. 32.
9. Quoted in Adrian Berry, *The Book of Scientific Anecdotes.* New York: Prometheus Books, 1993, pp. 173–74.
10. Quoted in James Gleick, *Genius: The Life and Science of Richard Feynman.* New York: Pantheon, 1992, pp. 155–56.
11. Quoted in Gleick, *Genius*, p. 156.

## Chapter 2: The Expanding Universe

12. Quoted in John Carey, ed., *The Faber Book of Science.* London: Faber and Faber, 1995, p. 416.

## Chapter 3: The Code of Life

13. Quoted in N. A. Tiley, *Discovering DNA.* New York: Van Nostrand Reinhold, 1983, p. 244.
14. Quoted in Lisa Yount, *Twentieth-Century Women Scientists.* New York: Facts On File, 1996, p. 69.
15. Quoted in Joseph Schwartz, *The Creative Moment: How Science Made Itself Alien to Modern Culture.* New York: HarperCollins, 1992, p. 123.
16. Quoted in Lisa Yount, *Cancer.* San Diego: Lucent Books, 1991, p. 30.
17. Richard Golob and Eric Brus, eds., *The Almanac of Science and Technology.* Boston: Harcourt Brace Jovanovich, 1990, p. 90.
18. Joel Davis, "Interview: Leroy Hood," *Omni*, November 1987, p. 118.
19. Nigel Calder and John Newell, eds., *On the Frontiers of Science.* New York: Facts On File, 1989, p. 120.

## Chapter 4: Magic Bullets

20. Quoted in Donald Robinson, *The Miracle Finders.* New York: David McKay, 1976, p. 3.
21. Quoted in Jenny Sutcliffe and Nancy Duin, *A History of Medicine.* New York: Barnes and Noble, 1992, p. 137.
22. Quoted in Edward Shorter, *The Health Century.* New York: Doubleday, 1987, p. 186.
23. Quoted in William Hoffman and Jerry Shields, *Doctors on the New Frontier.* New York: Macmillan, 1981, pp. 90–91.
24. Quoted in Hoffman and Shields, *Doctors on the New Frontier*, p. 91.
25. Quoted in Hoffman and Shields, *Doctors on the New Frontier*, pp. 95–96.

26. Quoted in Hoffman and Shields, *Doctors on the New Frontier*, p. 96.
27. Quoted in Robinson, *The Miracle Finders*, p. 233.
28. Quoted in Shorter, *The Health Century*, p. 125.
29. Quoted in Lisa M. Krieger, "AIDS Fight Shifts Focus—to Life," *San Francisco Examiner*, August 4, 1996.

## Chapter 5: Rebuilding the Body

30. Quoted in Catherine Caufield, *Multiple Exposures*. London: Secker and Warburg, 1989, p. 4.
31. Quoted in Jack Fincher, "New Machines May Soon Replace the Doctor's Black Bag," *Smithsonian*, January 1984, p. 67.
32. Quoted in Howard Sochurek, "Medicine's New Vision," *National Geographic*, January 1987, p. 19.
33. Quoted in Joan Morris, "A Cut Above," *West County Times*, January 4, 1991.
34. "Robot 'Surgeon' Operates on the Brain," *San Francisco Chronicle*, June 25, 1985.
35. Quoted in Robinson, *The Miracle Finders*, pp. 70–71.
36. Quoted in Robinson, *The Miracle Finders*, p. 71.
37. Quoted in Sherry Baker, "Internal Medicine," *Omni*, January 1991, p. 42.

## Chapter 6: The Information Revolution

38. Quoted in Stan Augarten, *Bit by Bit: An Illustrated History of Computers*. New York: Ticknor & Fields, 1984, p. 135.
39. Quoted in Clifford Stoll, *Silicon Snake Oil*. New York: Anchor Books, 1995, p. 12.
40. Quoted in Dennis Shasha and Cathy Lazere, *Out of Their Minds: The Lives and Discoveries of 15 Great Computer Scientists*. New York: Springer-Verlag, 1995, pp. 24–25.
41. Quoted in Shasha and Lazere, *Out of Their Minds*, p. 32.
42. Daniel Crevier, *AI: The Tumultuous History of the Search for Artificial Intelligence*. New York: Basic Books, 1993, p. 153.

## Chapter 7: An Unpredictable Future

43. Quoted in "1984: If This Is the Future . . . ," *Science 84*, January/February 1984, p. 39.
44. Quoted in "1984: If This Is the Future . . . ," p. 39.
45. Quoted in James Gleick, *Chaos: Making a New Science*. New York: Penguin Books, 1987, p. 14.
46. Quoted in Fritjof Capra, *The Turning Point*. New York: Simon and Schuster, 1982, p. 140.
47. Quoted in Timothy Ferris, ed., *The World Treasury of Physics, Astronomy, and Mathematics*. Boston: Little, Brown, 1991, p. 101.
48. Quoted in Alan Mackay, ed., *Scientific Quotations: The Harvest of a Quiet Eye*. New York: Crane, Russak, 1977, p. 72.
49. Quoted in Eugene P. Odum, *Ecology and Our Endangered Life-Support Systems*. Sunderland, MA: Sinauer Associates, 1993, p. 279.
50. Robert Heinlein, *Expanded Universe*, New York: Ace Books, 1982, p. 323.

# Glossary

**acquired immunodeficiency syndrome (AIDS):** A syndrome associated with the destruction of the immune system, generally thought to be caused by the human immunodeficiency virus (HIV).

**African Eve theory:** The theory that modern human beings first evolved in Africa and then spread to other continents.

**alternating current:** An electric current that constantly changes in charge and direction of flow.

**amino acid:** One of twenty types of small molecules that combine to make proteins.

**Andromeda:** A nearby galaxy, the first one outside the Milky Way to be identified.

**anode:** A positively charged conductor, or electrode.

**antibiotic:** A drug made from a substance produced by one microbe to kill another.

**antibody:** A chemical in the blood that reacts to antigens foreign to the body.

**antigen:** A substance on the surface of cells, to which the immune system may react.

**anti-oncogene:** A gene that turns off cell growth; if it is missing or damaged, cancer may result.

**artificial intelligence (AI):** The attempt to create computer programs that can reason or otherwise show intelligent behavior.

**atomic number:** The number of protons in an atom of an element.

**atomic weight:** The number of protons and neutrons in the nucleus of an element's atoms.

**azidothymidine (AZT):** A drug used to slow the progress of AIDS.

**base:** One of four chemicals that form the cross-links in a molecule of nucleic acid; their order determines the "code" through which inherited information is translated into protein.

**Big Bang:** The explosion believed to have created the universe about fifteen billion years ago.

**black hole:** A star so dense that not even light can escape its gravity.

**blood type:** A grouping that depends on the types of antigens on blood cells and determines the success of blood transfusions.

**cancer:** A class of diseases in which certain body cells multiply uncontrollably, damaging the rest of the body.

**cathode:** A negatively charged conductor (electrode) that can give off streams of electrons (cathode rays) in a vacuum tube.

**cell:** The self-contained, microscopic unit of which the bodies of living things are made.

**cepheid:** A variable star that goes from dim to bright to dim again in a regular cycle.

**chain reaction:** Splitting of atoms so that they release neutrons that in turn split more atoms.

**chaos theory:** A theory stating that the details and course of nature will sometimes be unpredictable.

**chromosome:** Threadlike body in the nucleus of a cell that carries material containing genes (DNA).

**COBOL (Common Business-Oriented Language):** An English-like computer programming language.

**compiler:** A program that translates instructions in a computer programming language into commands that can be executed by the computer.

**computerized axial tomography (CT or CAT):** A diagnostic technology that combines computers and X rays to create pictures of the inside of the body.

**contagious disease:** A disease that can be spread from one person to another.

**Crookes tube:** A glass tube from which the air has been removed, containing two electrodes through which a current can be passed.

**cyclosporine:** A drug that suppresses the immune system and keeps the body from rejecting transplanted organs.

**cytoplasm:** The jellylike material that makes up the body of a living cell.

**data:** Pieces of information, especially those used by a computer.

**deoxyribonucleic acid (DNA):** The chemical that contains, in coded form, the information that is inherited from parents and that controls a cell's activities.

**diagnosis:** Finding out the cause and nature of a medical condition.

**dialysis:** A process that filters wastes from the blood artificially, taking the place of a kidney.

**direct current:** An electric current that flows in a single direction.

**dominant trait:** A trait that will appear even if a living thing inherits a gene for it from only one parent.

**Doppler shift:** The power of waves of light or sound to bunch up as they approach an observer and stretch out as they move farther away.

**electroencephalograph (EEG):** A device that helps doctors diagnose problems in the brain by picking up electric signals that the brain gives off.

**electron:** A negatively charged particle found outside an atom's nucleus.

**electronics:** The technology of precisely controlling the flow of electromagnetic energy.

**element:** A substance that cannot be broken down into other substances.

**endoscope:** A tube that can be inserted into the body to allow doctors to look inside the body.

**evolution:** The theory that, over long periods of time, living things change through the mechanism of heredity (genetics) to become better adapted to their environment.

**expert system:** A computer program that applies a set of rules to draw conclusions about situations presented to it.

**fertilized egg:** The single cell from which a new living thing will be made; it is formed by the union of an egg from the mother and a sperm from the father.

**fiber optics:** A technology that uses flexible strands of glass to conduct light.

**fission:** Process by which a neutron strikes the nucleus of an atom and splits it into two parts, producing energy and additional neutrons.

**fusion:** Process by which two atoms come together to form a new atom, releasing a great amount of energy.

**galaxy:** A large system of stars held together by gravity.

**gene:** A stretch of bases on a DNA molecule that contains the code for one protein.

**gene therapy:** Treating disease by replacing damaged genes with healthy ones.

**genetic engineering:** A field of technology in which genes are artificially changed or combined.

**genetics:** The study of the way living things inherit and are shaped by information from their parents.

**genome:** The collection of all the genes in a cell.

**geology:** The study of the earth.

**helix:** A coiled shape; the DNA molecule consists of two such shapes twined around each other.

**human immunodeficiency virus (HIV):** Virus that most researchers believe to be the cause of AIDS.

**hypertext:** A way to link documents so that selecting highlighted text or pictures on one document leads the viewer to related documents.

**immune system:** The internal system, consisting of certain cells and chemicals, that defends the body against attack by microbes or other foreign substances.

**integrated circuit:** A group of electronic components and circuits placed on a single piece of material, such as a silicon wafer.

**Internet:** The loosely organized worldwide system of computer networks that allows users to exchange information.

**ion:** An atom to which electrons have been added or removed, so that it carries an electric charge.

**iproniazid:** A drug originally used to treat tuberculosis that proved useful in treating depression.

**laparoscope:** A type of fiber-optic endoscope used in abdominal surgery.

**laser:** A tool that uses a concentrated beam of light in which all the waves have the same length and move in the same direction.

**light-year:** The distance light travels in one year; about 5.9 trillion miles.

**magnetic resonance imaging (MRI):** A diagnostic technology that uses signals sent out by atoms in a magnetic field to make pictures of the inside of the body.

**mantle:** Layer of very hot liquid rock below the earth's crust.

**meteorology:** The study of weather and climate.

**microbe:** A living thing too small to see without a microscope; a microorganism.

**microcomputer:** A desk-size or smaller computer that features a microprocessor.

**microprocessor:** A silicon wafer or "chip" that contains a complete central processing unit (CPU).

**Milky Way:** The galaxy of which the solar system is a part.

**mitochondria:** Small bodies inside cells; they are outside the nucleus, but they contain DNA.

**molecular biology:** The study of the chemistry of cells, especially of genes, and the way that chemistry affects living things.

**multimedia:** A technology that allows video and sound to be combined with text in a presentation.

**mutation:** A change in a gene.

**natural selection:** According to Charles Darwin's theory of evolution, the process by which the environment determines the survivability of the traits inherited by living things.

**neural network:** A computer program that simulates the way brain cells respond to stimuli and learn.

**neutron:** A neutral (uncharged) particle found in the nucleus of atoms.

**nitrogen mustard:** The first chemical used to treat a variety of cancers; a relative of poisonous mustard gas.

**nuclear power:** Energy from controlled fission or fusion of atomic nuclei.

**nucleus:** The central part of an atom or of a cell.

**object:** In computer programming, a way to package related data and operations.

**oncogene:** A growth-producing gene that has mutated in a way that makes it cause a cell to multiply uncontrollably, producing cancer.

**parallax:** The difference in angle of an object as seen from two different locations; used to measure distance.

**penicillin:** An antibiotic made from the mold *Penicillium notatum.*

**pharming:** Using genetic engineering to make the bodies of farm animals produce substances useful in medicine.

**pitchblende:** A mineral containing uranium and other radioactive elements.

**plate tectonics:** Theory stating that the earth's crust consists of separate plates that move, carried by currents in the underlying mantle.

**polio:** Virus-caused contagious disease that frequently leaves its victims paralyzed; also called infantile paralysis.

**protein:** A type of chemical through which most activities in living cells are carried out.

**proton:** A positively charged particle found in the nucleus of atoms.

**quantum theory:** A theory that says that energy can be transferred between particles such as electrons only in discrete units called quanta.

**quark:** One of six fundamental particles from which other atomic particles such as protons and neutrons are formed.

**radioactivity:** Action by which certain atoms spontaneously break up and release particles or radiation.

**radium:** A radioactive element discovered by Marie and Pierre Curie.

**recessive trait:** A trait that will appear only if a living thing inherits genes for it from both parents.

**reductionism:** The belief that anything in nature can be explained by a small number of natural laws.

**reserpine:** A calming or tranquilizing drug made from a plant called snakeroot; one of the first drugs to successfully treat severe mental illness.

**ribonucleic acid (RNA):** A nucleic acid similar to DNA, used to help cells make proteins following DNA instructions.

**schizophrenia:** A severe mental illness usually marked by lack of contact with the outside world and the presence of delusions (false beliefs) and hallucinations (false experiences of sights or sounds).

**sex cell:** A cell that carries genetic information from a living thing to its offspring; an egg (female) or a sperm (male).

**shell:** A particular orbit in which electrons are held by the laws of quantum mechanics.

**simulation:** A computer program that imitates a real or imagined situation.

**singularity:** A condition inside a black hole in which normal laws of physics do not apply.

**sonography (ultrasound):** A diagnostic technology that uses very-high-frequency sound waves to make pictures of the inside of the body.

**spectrum:** A series of bands or lines into which light can be broken.

**subroutine:** A part of a computer program that can be called upon to carry out a particular task.

**sulfanilamide:** A substance from which a family of microbe-controlling drugs was made in the 1930s.

**supernova:** A gigantic explosion that marks the end of the life cycle of some stars.

**technology:** The application of science to meet human needs.

**trait:** A characteristic of a living thing, such as eye color.

**vaccine:** A substance, consisting of killed or weakened disease-causing microbes, that makes the immune system ready to fight a certain type of microbe.

**Van Allen belt:** Belt of highly charged atomic particles that circle the earth.

**World Wide Web:** A facility allowing people to create, display, and browse hypertext documents as well as providing access to pictures, data files, and so on.

**X rays:** A type of radiation that can pass through soft substances such as body tissue.

# For Further Reading

Isaac Asimov, *Asimov's New Guide to Science*. New York: Basic Books, 1984. A comprehensive and readable one-volume description of essential concepts and discoveries in every field of modern science.

Stan Augarten, *Bit by Bit: An Illustrated History of Computers*. New York: Ticknor & Fields, 1984. Well-illustrated account of early computers and technical developments.

Lennard Bickel, *The Deadly Element: The Story of Uranium*. New York: Stein and Day, 1979. The story of the discovery of radiation and nuclear energy, told in vivid and accessible language.

Charlene Billings, *Grace Hopper: Navy Admiral and Computer Pioneer*. Hillside, NJ: Enslow, 1989. The story of a remarkable woman who succeeded in two male-dominated worlds: the military and computer science. Written for young adults.

Nigel Calder and John Newell, eds., *On the Frontiers of Science*. New York: Facts On File, 1989. Well-illustrated book describes frontiers of modern science and predicts scientific advances of the future.

Edward Edelson, *Genetics and Heredity*. New York: Chelsea House, 1990. Overview of genetics for young adults.

Sean M. Grady, *Marie Curie*. San Diego: Lucent Books, 1992. Biography of Marie Curie for young adults.

Stephen Hawking, *A Brief History of Time: from the Big Bang to Black Holes*. New York: Bantam Books, 1990. Hawking explains theories about black holes and the ultimate nature of the universe. Not easy reading, but rewarding.

"1984: If This Is the Future . . . ," *Science 84*, January/February 1984, pp. 34–43. Amusing collection of predictions about twentieth-century technology.

Dennis Overbye, *Lonely Hearts of the Cosmos: The Scientific Quest for the Secret of the Universe*. New York: HarperCollins, 1991. Very readable account that brings out the human side of modern astronomy; describes the key discoveries about the expanding universe and its possible future.

Donald Robinson, *The Miracle Finders*. New York: David McKay, 1976. Vividly describes the stories behind twentieth-century medical advances.

Edward Shorter, *The Health Century*. New York: Doubleday, 1987. Interesting account of advances in twentieth-century American medicine, focusing on the government-sponsored National Institutes of Health.

Howard Sochurek, "Medicine's New Vision," *National Geographic*, January 1987, pp. 2–40. Describes CT, MRI, and other new imaging tools that help doctors see inside the body and diagnose disease.

James D. Watson, *The Double Helix*. New York: Atheneum, 1968. Lively, if biased, memoir of the discovery of the structure of DNA.

Trevor I. Williams, *Science: A History of Discovery in the Twentieth Century*. Oxford: Oxford University Press, 1990. Well-illustrated history of twentieth-century science and technology.

# Works Consulted

Frederick Lewis Allen, *The Big Change*. New York: Harper and Brothers, 1952. Fascinating description of the ways life in the United States changed between 1900 and 1950.

Sherry Baker, "Internal Medicine," *Omni*, January 1991, pp. 41–42, 79–80. Describes possibility of growing new human organs to replace damaged ones.

Adrian Berry, *The Book of Scientific Anecdotes*. New York: Prometheus Books, 1993. A collection of stories about scientific discoveries.

Fritjof Capra, *The Turning Point*. New York: Simon and Schuster, 1982. Interesting criticism of reductionism in twentieth-century science.

John Carey, ed., *The Faber Book of Science*. London: Faber and Faber, 1995. Collection of essays on a variety of science topics.

Catherine Caufield, *Multiple Exposures*. London: Secker and Warburg, 1989. Description of the impact that discoveries related to radioactivity have had on society.

*The Colonial Overlords: TimeFrame AD 1850–1900*. Alexandria, VA: Time-Life, 1990. Well-illustrated description of several aspects of world history during the second half of the nineteenth century.

Daniel Crevier, *AI: The Tumultuous Search for Artificial Intelligence*. New York: Basic Books, 1993. Describes development of the field from the 1950s through the 1980s.

Joel Davis, "Interview: Leroy Hood," *Omni*, November 1987, pp. 116–24, 150–56. Interview with the man who invented a machine that finds the base sequence in genes automatically.

Richard Dawkins, *River Out of Eden: A Darwinian View of Life*. New York: Basic Books, 1996. Interesting account of evolution based on the idea that the main purpose of living things is to carry and pass on their genes.

Bernard Dixon, ed., *From Creation to Chaos: Classic Writings in Science*. London: Basil Blackwell, 1989. Collection of writings that illuminate scientific discoveries and ideas.

Timothy Ferris, ed., *The World Treasury of Physics, Astronomy, and Mathematics*. Boston: Little, Brown, 1991. Collection of excerpts in which scientists describe their own discoveries.

Jack Fincher, "New Machines May Soon Replace the Doctor's Black Bag," *Smithsonian*, January 1984, pp. 64–70. Article describes CT, MRI, and other new imaging tools that help doctors diagnose disease.

*Genetics: Readings from* Scientific American. San Francisco: W. H. Freeman, 1981. Key papers on genetics, including those of Mendel and Francis Crick; difficult reading.

James Gleick, *Chaos: Making a New Science*. New York: Penguin Books, 1987. Describes the discovery of unpredictable, chaotic phenomena and chaos theory's impact on the way scientists see the world.

———, *Genius: The Life and Science of Richard Feynman*. New York: Pantheon, 1992. An interesting, readable biography of a brilliant and quirky physicist involved in projects ranging from the atomic bomb to the investigation of the *Challenger* disaster.

Richard Golob and Eric Brus, eds., *The Almanac of Science and Technology*. Boston: Harcourt Brace Jovanovich, 1990. Excellent, if somewhat difficult, review of current knowledge in a variety of sciences.

Robert Heinlein, *Expanded Universe*. New York: Ace Books, 1982. Collection of stories and essays by a pioneer science fiction writer; some material dated, but interesting nevertheless.

Harry Henderson, *Stephen Hawking*. San Diego, CA: Lucent Books, 1995. Biography of Hawking for young people; tells of his work with black holes and his struggle to overcome his disability.

William Hoffman and Jerry Shields, *Doctors on the New Frontier*. New York: Macmillan, 1981. Presents the stories behind several recent medical discoveries, including Nathan Kline's antipsychotic drugs.

Lisa M. Krieger, "AIDS Fight Shifts Focus—to Life," *San Francisco Examiner*, August 4, 1996. Describes more hopeful attitudes of patients and doctors brought about by improved treatments for AIDS.

Alan Mackay, ed., *Scientific Quotations: The Harvest of a Quiet Eye*. New York: Crane, Russak, 1977. A collection of short quotes by scientists, organized alphabetically.

Douglas McKie, *Antoine Lavoisier: Scientist, Economist, Social Reformer*. New York: Henry Schuman, 1952. Biography of the eighteenth-century French scientist often called the father of modern chemistry.

Joan Morris, "A Cut Above," *West County Times*, February 4, 1991. Describes how laparoscopes and lasers have improved abdominal surgery.

John Naisbitt and Patricia Aburdene, *Megatrends 2000: New Directions for the 1990's*. New York: William Morrow, 1990. A popular discussion of technological and social trends in the years approaching the twenty-first century.

Eugene P. Odum, *Ecology and Our Endangered Life-Support Systems*. Sunderland, MA: Sinauer Associates, 1993. College text discussing ecology and its relationship to human society.

"Robot 'Surgeon' Operates on the Brain," *San Francisco Chronicle*, June 25, 1985. Describes surgeon Yik San Kwoh, who uses a robot to assist him in operating on the brain.

Carl Sagan, *The Demon-Haunted World: Science as a Candle in the Dark*. New York: Random House, 1995. Using incidents from his own life, the famous astronomer-author tries to show that science is beautiful and meaningful. Sagan also debunks a variety of pseudoscientific beliefs.

Joseph Schwartz, *The Creative Moment: How Science Made Itself Alien to Modern Culture.* New York: HarperCollins, 1992. Describes important discoveries in twentieth-century science and how they affected society.

Dennis Shasha and Cathy Lazere, *Out of Their Minds: The Lives and Discoveries of 15 Great Computer Scientists.* New York: Springer-Verlag, 1995. A good but rather technical look at key computer scientists and their work.

Jane S. Smith, *Patenting the Sun.* New York: William Morrow, 1990. Detailed description of the development of the Salk polio vaccine.

Clifford Stoll, *Silicon Snake Oil.* New York: Anchor Books, 1995. A skeptical view of the Internet; a thought-provoking perspective on the "information superhighway."

Jenny Sutcliffe and Nancy Duin, *A History of Medicine.* New York: Barnes and Noble, 1992. Readable, well-illustrated overview of medicine, focusing on the twentieth century.

Lewis Thomas, *The Lives of a Cell.* New York: Viking, 1974. Beautifully written essays on various aspects of biology and medicine.

N. A. Tiley, *Discovering DNA.* New York: Van Nostrand Reinhold, 1983. History of genetics and genetic engineering; difficult reading but contains interesting quotes.

Peter Vansittart, ed., *Voices 1870–1914.* New York: Franklin Watts, 1985. Anthology of short quotations that reveal the mood of the time.

Lael Wertenbaker, *To Mend the Heart.* New York: Viking Press, 1980. Describes the dramatic stories of heart surgery pioneers.

Lisa Yount, *Cancer.* San Diego: Lucent Books, 1991. Overview of recent research on the nature, causes, and treatment of cancer.

———, *Twentieth-Century Women Scientists.* New York: Facts On File, 1996. Biographies of women scientists from around the world, including Rosalind Franklin, whose work contributed to understanding the structure of DNA.

# Index

ADA (adenosine deaminase), 52–53
African Eve theory, 48
AI (artificial intelligence), 77, 86–88
AIDS (acquired immunodeficiency syndrome), 64–65
alpha particles, 20
alternating current, 78
amino acids, 46–47
Anderson, W. French, 53
Andromeda nebula, 27, 28, 32
anodes, 16–17, 78
antibiosis, 55
antibiotics, 56, 57, 58, 64
antibodies, 74
antigens, 74
anti-oncogenes, 49
Apple Computer Company, 83–84
Armstrong, Edwin, 79
artificial body organs, 72–73
artificial intelligence (AI), 77, 86–88
atomic bombs, 12, 21–22, 23
atoms, 14, 16–18
    composition of, 20
    current research on, 25
    electron orbits in, 18–19
    splitting of, 20–21
AZT (azidothymidine), 65

bacteria, 50–51, 54
    penicillin and, 55
Balfour, Lord Arthur, 10
Bargar, William, 72
barium, 67
Barnard, Christiaan, 76
BASIC computer language, 84

Becquerel, Antoine–Henri, 14–15
Becquerel rays, 15
Bell, Alexander Graham, 90
Berg, Paul, 50
beryllium, 20
Bessel, Friedrich Wilhelm, 26
Bethe, Hans, 29
Big Bang theory, 33, 34–35
biotechnology, 50–52
Bishop, Michael, 49
black hole singularity, 33–34
Blaese, R. Michael, 52
Blaiberg, Philip, 76
blood
    transfusions, 74–75
    types, 74
Bohr, Niels, 19
Boyer, Herbert, 50
Bozzini, Phillipe, 67
Braun, Wernher von, 89

cancer, 49, 64, 75
    nitrogen mustard and, 59–60
Cannon, Annie J., 28
cathodes, 16, 17, 77, 78
CAT scans, 68, 69
cells
    cytoplasm of, 46
    division of, 42–43
    gene mutation and, 48–50
    nucleus of, 42
cepheids, 26–27
CERN (Swiss physics institute), 85
Chadwick, James, 20
Chain, Ernst, 55–56
Chandrasekhar, Subrahmanyan, 33
chaos theory, 94
Chernobyl nuclear accident, 24
chlorpromazine, 63, 64
chromosomes, 47
    cell division and, 42–44
    duplication of, 44–46
Churchill, Winston, 23
Clarke, Arthur C., 79
Clarke, Barney, 73
COBOL (Common Business-Oriented Language), 82
Cohen, Stanley, 50
coherer (early radio), 77
computers, 71, 86
    bugs in, 80
    diagnostic technology and, 67, 68
    global communication and, 79
    languages for, 81–82, 84
    mini, 83
    natural laws and, 92–93
    programs for, 77, 81–83, 86–88
    surgery and, 71, 72
    telescopes and, 31
continental drift, 37–39
Crick, Francis, 44–47
Crookes tube, 16, 17, 78
Curie, Eve, 18
Curie, Marie, 15–16, 18
Curie, Pierre, 15, 18
cyclosporine, 76
cytoplasm, 46, 48

Darvall, Denise, 75–76
Darwin, Charles, 42
ddI (didanosine), 65
depression, 60, 63–64
DeSilva, Ashanthi, 52–53
dialysis machines, 72–73
direct current, 78

DNA (deoxyribonucleic acid), 65
   genetic clocks and, 48
   genetic engineering and, 50–51
   structure of, 44–46, 47
Doherty, Thomas, 60
Domagk, Gerhard, 54–55
dominant traits, 41
Donald, Ian, 68
Doppler, Christian, 32
drug companies, 57, 61–63
drugs
   depression, 63–64
   HIV, 65
   penicillin, 55–56
   polio, 58–60
   schizophrenia, 60–63
   sulfa, 54

Eckert, John Presper, 80–81
EDVAC computer, 81
EEG (electroencephalograph), 67
Ehrlich, Paul, 54
Einstein, Albert, 21–22, 34
electricity
   charges of, 16–17, 25
   currents of, 67, 78
electrocardiograph, 67
electroencephalograph (EEG), 67
electromagnetic signals, 68, 69, 77–79
electronic mail, 85
Electronic Numerical Integrator and Calculator (ENIAC), 80–81
electronics, 77–79
electrons, 16–17
   orbits of, 18–19
   protons and, 20
   radio signals and, 77–78
elements, 15–16, 18–19, 20–21
e-mail, 85

Enders, John, 58
endoscopes, 67–68, 69, 70
Englebart, Douglas, 85
ENIAC (Electronic Numerical Integrator and Calculator), 80–81
Ensor, James, 10
evolution, 42, 48
expert systems, 87–88
*Explorer I*, 35
eye surgery, 69–70

Fermi, Enrico, 20–21
Feynman, Richard, 22
fiber optics, 68, 70
fission, atomic, 21, 22–23
Fleming, Alexander, 55
Fleming, Ambrose, 78
Florey, Howard, 55–56
Forest, Lee de, 78–79, 89
Frederickson, Donald, 91
Frisch, Otto, 21
fusion, nuclear, 24–25, 29

galaxies, 27, 28
   Andromeda nebula, 27, 28, 32
   movement of, 32–33
Gamow, George, 33
Gates, Bill, 84
Gell–Mann, Murray, 25
genes, 44, 46
   evolution and, 48
   genetic engineering and, 50–51
   Human Genome Project and, 49–50
   inherited traits, 40
   mutations in, 48–49, 52–53
   production of protein in, 46–47
gene therapy, 52–53
genetic clocks, 48
genetic engineering, 50–53
   "organoids" and, 76
globular clusters, 27

Hahn, Otto, 20–21
Hawking, Stephen, 34
heart
   machines, 73
   transplants, 75–76
Heisenberg, Werner, 91–92, 94
helium, 28, 29–30
helixes, 44, 45, 47
Helmholtz, Hermann von, 29
Hernandez, Sandra, 65
Hertzsprung, Ejnar, 28
Hertzsprung-Russell diagram, 28
Hewlett-Packard company, 83–84
HIV (human immunodeficiency virus), 65
home pages, 85–86
Hood, Leroy, 50
Hopfield, John, 88
Hopper, Grace, 80, 82
Hounsfield, Geoffrey, 68
Hoyle, Fred, 29–30, 33
Hubble, Edwin, 27, 28, 32–33
Hubble space telescope, 36–37
Human Genome Project, 49–50
human immunodeficiency virus (HIV), 65
hydrogen, 28, 29, 32
hypertext, 85

IBM (International Business Machines), 81, 84
immune system
   ADA and, 52–53
   blood transfusions and, 74
   cancer and, 59–60
   HIV and, 65
   transplanted organs and, 75, 76
   vaccines and, 56–57
inherited diseases, 52–53

inherited traits, 40
Internet, 85–86
ions, 16–17
iproniazid, 63–64

Jarvik, Robert, 73
Jobs, Steve, 83–84
Jupiter, 36, 37

Kelvin, Lord, 14, 29
kidney
    machines, 72–73
    transplants, 75, 76
Kline, Nathan, 60–61, 63–64
Kolff, Willem, 72–73
Kwoh, Yik San, 71–72

Landsteiner, Karl, 74–75
laparoscopes, 70–71
LaPlace, Pierre de, 91
lasers, 69–70
Lasker Award, 64
Leavitt, Henrietta, 26–27
leukemia, 60
light-years, 26, 27, 28
Lorenz, Edward, 93–94
Lowell, Percival, 35
lymphoma, 59–60

magnetic resonance imaging (MRI), 68–69
March of Dimes, 57–58
Marconi, Guglielmo, 77–78
*Mariner 2*, 35
Mark II computer, 80
Mars, 35–36, 89
Mauchly, John, 80–81
McCarthy, John, 86–87
Meitner, Lise, 21
Mendel, Gregor, 40–42
mental hospitals, 62, 64
mental illnesses, 60–64
microbes, 54, 55, 56, 65, 75
    in vaccines, 57
    resistant, 64
microchip, 83

microcomputers, 83–84
microprocessors, 83
Microsoft company, 84
Milky Way, 27, 28, 32
molecular biology, 47
molecules, 44–46
MRI (magnetic resonance imaging), 68–69
multimedia programs, 85

NASA (National Aeronautics and Space Administration), 35
National Foundation for Infantile Paralysis (NFIP), 57–58, 59
*Nautilus* (submarine), 22
nebulae, 27
Nelson, Ted, 85
neural networks, 88
neutrons, 20–21
    in stars, 33
    nuclear reactors and, 23–24, 25
Newton, Isaac, 91, 94
NFIP (National Foundation for Infantile Paralysis), 57–58, 59
nitrogen mustard, 59–60
nuclear energy, 21, 22–24
    safety concerns of, 24–25
nuclear fusion, 24–25
nucleic acids, 44

oncogenes, 49
Oppenheimer, Robert, 22
organs
    replacement of, 72–73
    organ transplants, 75–76

parallax, 26
penicillin, 55–56, 57
pharming, 51
pitchblende, 14–15
Planck, Max, 19
plate tectonics, 38–39

Pluto, 36
polio, 57–59
polonium, 15
Prontosil, 54–55
protease inhibitors, 65
proteins, 44, 46
    gene mutations and, 48–49
    genetic engineering and, 50–51
protons, 20, 25
psychotherapy, 63, 64
Ptolemy, Claudius, 26

quantum theory, 19
quarks, 25

radiation, 14, 20, 32, 66
    Big Bang and, 33
radioactivity
    Curies and, 15–16
    nuclear power and, 22–24
radio waves, 77–79
    Big Bang and, 33
    MRI and, 68, 69
radium, 15, 18
recessive traits, 41
reductionism, 91, 93–94
reserpine, 61, 63, 64
RNA (ribonucleic acid), 46–47, 65
Robbins, Frederick, 58
robots, 35, 71–72
Röntgen, Wilhelm, 14, 16, 66
Roosevelt, Franklin D., 21–22, 57–58
Russell, Henry, 28
Rutherford, Ernest, 16, 17, 20

Sabin, Albert, 58
Sabin vaccine, 59
Salk, Jonas, 58, 61
Salk vaccine, 58–59, 61
satellites, 27, 35, 37, 79
schizophrenia, 60, 61, 63
Schroeder, William, 73
Scott, Charles P., 63

sex cells, 43–44
Shapley, Harlow, 27
Shumway, Norman, 76
Smith, John, 57
snakeroot, 60–61
solar system, 12, 26, 27
sonar technology, 68
sound waves, 32, 68
Soviet Union
    nuclear power and, 22
        Chernobyl, 24
    Sabin vaccine and, 59
    *Sputnik I* and, 35
space program, 35–37
spectrograph, 28, 29
*Sputnik I*, 35
Squibb company, 61–63
stars
    color of, 32
    Delta Cephei, 26
    distance of, 26–27
    fuel supply of, 29–30
    Magellanic Clouds, 26–27
    nuclear reactions on, 29–30
    Proxima Centauri, 26
    61 Cygni, 26
    spectra of, 28, 29
    temperature of, 28
    white dwarf, 30, 33
steady state theory, 33
streptococcus, 55–56
submarines, 22, 68
sulfa drugs, 54
sun, 28, 29–30

supergalaxies, 27
supernova, 30
surgical tools, 69–72
syphilis, 54

technology, 10
    diagnostic, 67–72
    history of, 12–13
    Human Genome Project and, 49–50
    mixed blessing of, 94–95
    predictions about, 89–91
    war and, 11, 12
    *see also* computers
telescopes, 26, 27, 31
Thompson, John, 76
Thomson, Joseph, 16–17, 77
thorium, 15, 16
Three Mile Island nuclear accident, 24
transplants, 72–76
    future of, 76
    immune systems and, 74

ultrasound, 68
ultraviolet light, 17
United States. *See* World War II
UNIVAC computer, 81, 82
universe, expansion of, 32–33, 34–35
uranium, 15–16, 20–21

vaccines, 56–59
    for HIV, 65

Vakil, Rustom, 60–61
Van Allen belt, 35
Varmus, Harold, 49
Venus, 35
Verne, Jules, 89
Viking project, 35–36
virtual reality, 71
viruses, 54
    contagious diseases and, 57, 58

Washkansky, Louis, 75–76
Watson, James, 44–47
Wegener, Alfred, 37–39
Weller, Thomas, 58
Wells, H. G., 11
Williams, Trevor I., 10, 13
World War I, 11, 12, 59
World War II, 12–13
    atomic bombs and, 22, 23
    blood transfusions during, 75
    computer development and, 80
    penicillin usage during, 55–56
    radio broadcasting industry and, 79
    sonar technology and, 68
World Wide Web, 85–86
Wozniak, Stephen, 83–84

X rays, 14–15, 66–68, 69

# Picture Credits

Cover photo: David Parker/Science Photo Library

© 1994, Baxter Healthcare Corp., 73

Agricultural Research Service, USDA, 53, 92

AIP Niels Bohr Library/Margrethe Bohr Collection, 19

American Cancer Society, 60

AP/Wide World Photos, 31, 32, 64, 87

Archive Photos, 50, 62, 63, 83

Corbis-Bettmann, 20

Library of Congress, 11, 12 (bottom), 15, 22, 58, 59 (both), 61, 90, 91

NASA, 13, 27, 28, 30 (bottom), 36, 37, 38, 79

Novosti Press Agency (APN), 24

*Otto Hahn: A Scientific Autobiography*. NY: Charles Scribners Sons, 1966. Courtesy AIP Niels Bohr Library, 21

Photo Researchers, Inc./© A. Barrington Brown/Science Source, 44

Photo Researchers, Inc./© Simon Fraser/ Science Photo Library, 74

Photo Researchers, Inc./Matt Meadows/ Science Photo Library, 67

Photo Researchers, Inc./© Hank Morgan/Science Source, 65, 75

Photo Researchers, Inc./© Omikron/ Science Source, 66

Photo Researchers, Inc./© Erich Schrempp, 33

Photo Researchers, Inc./© Siu/Science Source, 71

Photo Researchers, Inc./Wellcome Dept. of Cognitive Neurology/Science Photo Library, 69

Prints Old & Rare, 42

© Smithsonian Institution, 12 (top), 23, 78

University of Utah, 72

UPI/Bettmann, 55, 81, 82 (top)

UPI/Corbis-Bettmann, 30 (top), 39, 45, 56, 82 (bottom)

Woodfin Camp & Associates, 14

Woodfin Camp & Associates/Hulton Deutsch Collection Limited, 29, 40, 54

Woodfin Camp & Associates/Sipa Press, 84, 85, 93, 95

# About the Authors

Harry Henderson and Lisa Yount are a husband-and-wife team who live with a large library and four cats in El Cerrito, California. Harry Henderson is a technical writer who specializes in computer programming and communications. His books for young people include *Stephen Hawking* (Lucent Books) and *Modern Mathematicians* (Facts On File). Lisa Yount has written educational material for young people for over twenty-five years. Her books for young people include *Twentieth-Century Women Scientists* (Facts On File) and *Pesticides* (Lucent Books). Henderson and Yount also coauthored *The Scientific Revolution* (Lucent Books).

## DATE DUE

| MAY 2 5 20?? | | | |
|---|---|---|---|
| | | | |
| | | | |
| | | | |
| | | | |
| | | | |
| | | | |
| | | | |
| | | | |
| | | | |
| | | | |
| | | | |
| | | | |
| | | | |
| | | | |
| | | | |
| | | | |

Demco, Inc. 38-293